John Croumbie Brown

Glances at the Forests of Northern Europe

John Croumbie Brown

Glances at the Forests of Northern Europe

ISBN/EAN: 9783744662062

Printed in Europe, USA, Canada, Australia, Japan

Cover: Foto ©berggeist007 / pixelio.de

More available books at **www.hansebooks.com**

GLANCES AT THE FORESTS

OF

NORTHERN EUROPE.

GLANCES AT THE FORESTS

OF

NORTHERN EUROPE.

BY THE

REV. J. C. BROWN, LL.D.

LONDON:

J. AND W. RIDER, 14, BARTHOLOMEW CLOSE.

1879.

WORKS ON FORESTRY BY DR. BROWN.

Reboisement in France; or, Records of the Replanting of the Alps, the Cevennes, and the Pyrenees with Trees, Herbage, and Bush, with a view to arresting and preventing the destructive consequences of torrents : In which are given a *résumé* of Surrel's study of Alpine torrents, and of the literature of France relative to Alpine torrents, and remedial measures which have been proposed for adoption to prevent the disastrous consequences following from them ; translations of documents and enactments showing what legislative and executive measures have been taken by the Government of France in connection with *réboisement* as a remedial application against destructive torrents ; and details in regard to the past, present, and prospective aspects of the work. London : Henry S. King and Co. 1876.

Pine Plantations on Sand-Wastes in France : In which are detailed the appearances presented by the Landes of the Gironde before and after culture, and the Landes of La Sologne ; the legislation and literature of France in regard to the planting of the Landes with trees ; the characteristics of the sand-wastes ; the natural history, culture, and exploitation of the maritime pine and of the Scotch fir ; and the diseases and injurious influences to which the maritime pine is subject. Edinburgh : Oliver and Boyd. London : Simpkin, Marshall, and Co. 1878.

Forests and Moisture ; or, Effects of Forests on Humidity of Climate. In which are given details of phenomena of vegetation on which the meteorological effects of forests affecting the humidity of climate depend,—of the effects of forests on the humidity of the atmosphere, on the humidity of the ground, on marshes, on the moisture of a wide expanse of country, on the local rainfall, and on rivers,—and of the correspondence between the distribution of the rainfall and of forests,—the measure of correspondence between the distribution of the rainfall and that of forests,—the distribution of the rainfall dependent on geographical position, determined by the contour of a country,—the distribution of forests affected by the distribution of the rainfall,— and the local effects of forests on the distribution of the rainfall within the forest district. Edinburgh : Oliver and Boyd. London : Simpkin, Marshall, and Co. 1877.

The Schools of Forestry in Europe : a Plea for the Creation of a School of Forestry in Edinburgh. Edinburgh : Oliver and Boyd. 1877.

The School of Forestry in the Polytechnic School of Carls-rube. The School of Forestry in the Royal Wurtemburg Academy of Land and Forest Economy. The School of Forestry in the Escurial of Spain. The School of Forestry at Evois in Finland. Opinions of Continental Foresters and Professors of Forest Science on the location of a School of Forestry. A British School of Forestry : Review of Suggestions relative to its formation. London : J. and W. Rider. 1877.

On Schools of Forestry. Reprinted from Transactions of the Scottish Arboricultural Society. Edinburgh : M'Farlane and Erskine. 1877.

PREFACE.

For some years I held, along with the Chair of Botany in the South African College, Cape Town, the appointment of Government Botanist at the Cape of Good Hope. In 1865, in giving evidence before a select Committee of the Legislative Council, in answer to the queries, "Have you anything to recommend as the subject of forest economy ? Can you recommend any other mode of taking care of the forests than what is now pursued under the superintendents?" I said, amongst other things, "And thirdly, I recommend the procuring information in regard to the most approved measures of forest economy which are applicable to the management of forests in the colony, by commissioning some one acquainted with these forests to visit the forest schools of Germany, and if it be thought desirable, to visit also the forests in other parts of Europe, and report thereafter what is seen, or suggested by what is seen there applicable to the management of forests in this country, whether relating to matters connected with private enterprise or Government control."

After my return to Europe I addressed myself to supply the desideratum, and prepared exhaustive reports on Forest Science, Forest Legislation, and Forest Management in France, on the Forest Schools and Forest Administration in Germany, and on the meteorological effects of forests. Having been invited to spend last summer in St. Petersburg, I availed myself of the opportunity to verify and extend information I had previously collected in regard to forestry in Scandinavia,

Finland, and Russia, by procuring official documents and reports recommended to me by forest officials and Professors and Directors of Schools of Forestry who had honoured me with their friendship and correspondence. These documents have now been translated and embodied in reports which may be published at any time should this seem desirable. While this work was in progress these " Glances " appeared in successive numbers of the *Journal of Forestry*, each of them revealing the particular aspect of forestry presented for study by the country to which they were devoted. In Denmark may be studied the remains of forests of prehistoric times; in Norway, luxuriant forests managed by each proprietor as seemeth good in his own eyes ; in Sweden, sustained systematic endeavours to regulate the management of forests in accordance with the latest deliverances of modern science; in Finland, *Sartage* disappearing before the most advanced modern forest economy; and in Russia, *Jardinage* in the north merging into more scientific management in Central Russia and *Reboisement* or sylviculture in the south.

JOHN C. BROWN.

Haddington,
   April, 1879.

CONTENTS.

		PAGE
I. DENMARK	. . . . . .	9
II. NORWAY	. . . .	17
III. SWEDEN	. . . .	27
IV. FINLAND	. . . . . .	41
V. NORTHERN RUSSIA	. . . .	47

GLANCES AT THE FORESTS OF NORTHERN EUROPE.

I.—DENMARK.

THE opening chapter of these glances at forestry in the north of Europe is assigned to Denmark because the idea of preparing them was suggested in connection with a professional visit to the north of Europe, which I was invited, as a minister, to make last summer, and Denmark was my point of departure, after crossing the German Ocean or North Sea. I may further state, that in preparing these sketches I avail myself freely of information previously obtained on similar journeys and otherwise, as well as of what I gathered in the course of this trip.

Copenhagen, the capital of Denmark, may be reached by railway from Hamburg in fifteen hours, or it may be reached by steamer direct from Hull, Newcastle, or Leith, in about forty-eight hours. The railway journey from Hamburg will enable the traveller to see something of the country; but more may be seen by proceeding by rail and steam from Copenhagen to Frederikshavn and Gottenburg, and thence to Stockholm, the capital of Sweden.

In Copenhagen the traveller should not fail to see the Thorwaldsen Museum, the statuary by Thorwaldsen in the Fruekirk, or Church of our Lady ; and if he is interested in anthropology and ethnography, the museums of objects illustrative of these in the Prindsen Palace should also be visited.

It is some forty years since I first visited Copenhagen. I was proceeding from St. Petersburg to London by the *Sirius*—the first steamer which crossed the Atlantic, and the first steamer which made a voyage from Britain to the capital of Russia. Coal had to be taken in at Copenhagen, and six hours being required for this, I went with my wife to see the city. Meeting a " douce canny " man walking slowly with his hands behind his back, and an umbrella under his arm, I accosted him, and in broken German asked the way to some object we wished to ·see. To my surprise he exclaimed in broad Scotch, " Ae, Mr. Broon, who's a wi' ee ? " I was taken aback. He proved to be the skipper of a Leith vessel who had heard me preach in Leith, and whose wife was an old acquaintance and friend of our family, though I had never before met with him. To his kindness in acting as a guide we owed much. He had been detained in port by unfavourable winds.

About six miles from Copenhagen is the *Dyrehave*, or Deer-park, a beautiful forest of oaks and beeches, easily reached by steamboat or rail, being a favourite resort of many people in summer. But to the student of forest science, Denmark is more interesting from illustrations which it supplies of what *has been*, than of what *is*, in connection with forestry. So late as the eleventh century Jutland was described as *horrida sylvis*, but it has gradually lost the greater part of its woods. From the western coast of Schleswig they have disappeared; on the eastern coast alone are they to be seen, and even there they are but thinly scattered over the country.

The kingdom of Denmark is composed of several islands, together with a portion of the continent of Europe. Some of these islands are characterized by fertility, and Zealand, the largest, and that on which the capital stands, has the face of the country beautifully diversified with woods and lakes; Jutland, the continental peninsula, is more fertile still.

"The aspect of the Danish islands," says a writer of the last generation, "in general, is pleasant and cheerful, consisting of plains intersected by gentle hills, sometimes insulated and sometimes continuous, forming agreeable valleys. The heights, for the most part, are clothed with pasture or shaded with tufts of trees, whilst clear and azure lakes occasionally animate the scene. The province of Jutland presents a ruder aspect, but at the same time more varied and imposing, diversified with majestic forests, upland moors, and fertile pastures. Holstein and Sleswick are level and well-culti_vated countries, resembling England in their variety of hills, woods, rivulets, meadows, and corn-fields. The environs of Plien are distinguished for their picturesque, and those of Sleswick-town, Flensburg, and Apeurad for their romantic beauties."

From statements made in Reventlovs *Wirksomhed som Kongens Embedsmand og Statens Borger*, edited by Bergsoe, and cited by Marsh in "Man and Nature; or, the Earth, as modified by Human Action," it appears that—

"The felling of the woods on the Atlantic coast of Jutland has exposed the soil not only to drifting sands, but to sharp sea winds that have exerted a sensible deteriorating effect on the climate of that peninsula which has no mountains to serve at once as a barrier to the force of winds, and as a storehouse of moisture received by precipitation or condensed from atmospheric vapours."

In Schleswick-Holstein, and in Jutland, there are upwards of a thousand German square miles of dunes and sand plains—1,005 German, or 20,350 English square miles.

Besides suffering from the drifting of these sands, which may be attributed in a great measure to the destruction of the forests, and which a *reboisement* of the country might arrest, Denmark has suffered oftener than once from disastrous encroachments of the ocean; and in other ways extensive lagoons have been formed. Thus Denmark has come to supply many illustrations of the effects of vegetation, both herbaceous and arborescent, in filling up and drying up marshes in prolonged periods, extending back into pre-historic times. And the Danish language is rich in specific designations given to different kinds or forms of bogs and moors thus formed, whereby the pheno- mena may be more accurately studied. It was in connection with this that my attention was first directed to the physical geography of the country.

Referring to a work entitled *Om Möjligheten och Fördelen af allmänna Uppodlingar i Lappmarken*, by L. L. Læstadius, published in Stockholm 1824, Mr. Marsh remarks,—

"The English nomenclature of this geographical feature does not seem well settled. We have *bog, swamp, marsh, morass, moor, fen, turf-moss, peat-moss quagmire*, all of which, though sometimes more or less accurately discrimi- nated, are often used interchangeably, or are perhaps employed, each exclu- sively, in a particular district. In Sweden, where, especially in the Lappish provinces, this terraqueous formation is very extensive and important, the names of its different kinds are specific and exclusive in their application."

Illustrations are given ; and something similar may be alleged in regard to the nomenclature of these in Denmark. The student of forest science finds the study of the natural history of these bogs thus greatly facilitated.

In Denmark the tree which most extensively prevails at the present day is the beech. When remains are found at all in a decaying state under the soil, it is in the uppermost layer of peat-bog. As we go further down in the peat, remains of the oak are abundant, though the oak is, in the Denmark of to-day, almost entirely superseded by the beech. At a still lower level we come upon remains of the Scotch fir, and these in great abundance, though this is a tree which has never, within the range of historical times, been indigenous in the country.

In glancing at phenomena thus referred to, there arises first the question, When did all this take place ? A second question is, Whence came the seeds producing the successive crops of different kinds of trees ? And a third is, What light may be thrown upon the physical condition of the country at successive periods by these successive products ? Omitting the discussion of the second of these, so as to gain space, I address myself to the first, and shall briefly advert to the last question ere I close.

The hunters of the old Danish firwoods, and the fishermen who produced the mounds of shells which are found in Denmark, lived before the ages indicated by the employment of bronze and of iron in the manufacture of weapons, utensils, and ornaments ; but at what period of time they lived we have not as yet the means of determining. " We cannot," says a writer on the antiquity of man, " prove that the Danish firs, and subsequently the Danish oaks, took a long time for their successive disappearance. It may, in each case, have been 100,000 years, or it may have been ten times as long ; but it is not impossible that it may have been in a much shorter time ; but it is alleged that these Danish fishermen were not the primitive race of men, for their flint implements are polished, and belong to what antiquarians designate the Neolithic period, which was preceded, if not in Denmark, elsewhere in Europe, by what is designated the Palæolithic period, during which the stone implements were unpolished and comparatively rude. These have been traced, or attributed, to a very remote antiquity, about which geologists as yet can talk but vaguely of the chronological landmarks by which its dates may be indicated.

The elevation of part of Denmark above the level of the sea may have been a comparatively recent occurrence, viewed in relation to occurrences indicated by geological phenomena not very remote. Sir Charles Lyell, in his " Principles of Geology" (9 ed. chap. xxx.), shows that the upward movement now in progress in parts of Norway and Sweden extends throughout an area of about 1,000 miles north and south, and for an unknown distance east and west, the amount of elevation always increasing as we proceed towards the North Cape, where it is said to equal five feet in a century.

" If we could assume," he says, " that there had been an average rise of two and a half feet in each hundred years for the last fifty centuries, this would give an elevation of 125 feet in that period. In other words, it would follow that the shores, and a considerable area of the former land, of the North Sea had been uplifted vertically to that amount and converted into land in the course of the last 5,000 years."

Referring to certain post-tertiary marine deposits found at varying elevations up to 600 feet in Norway, where they are usually described as " raised beaches," he says,—

" A mean rate of continuous vertical elevation of two and a half feet in a century would, he considered, be a high average: yet even if this be assumed, it would require 24,000 years for these parts of the sea-coast of Norway to attain that elevation; these or cotemporaneous deposits contain shells,— different, indeed, from those of the brackish water character peculiar to the Baltic, but such as now lie in the Northern Ocean."

In the *Bulletins de la Société des Sciences*, t. vi. (Lausanne, 1860), there is a good account of the researches of Danish naturalists and antiquaries by an able Swiss geologist, M. A. Marlot. Apparently with this before him, Sir Charles Lyell, in his " Treatise on the Geological Evidences of the Antiquity of Man," writes (pp. 8, 9):—

" The deposits of peat in Denmark, varying in depth from ten to thirty feet, have been formed in deep hollows or depressions in the northern drift, or boulder formation. The lowest stratum, two or three feet thick, consists of swamp peat composed chiefly of moss, or sphagnum, above which lies another growth of peat, not made up exclusively of aquatic or swamp plants. Around the borders of the bogs, and at various depths in them, lie trunks of trees, especially of the Scotch fir (*Pinus sylvestris*), often three feet in diameter, which must have grown on the margin of the peat mosses, and have frequently fallen into them."

Other trees are mentioned by him as found at lesser depths; but what more concerns us here, he goes on to say,—

" All the land and fresh-water shells, and all the mammalia, as well as the plants whose remains occur buried in the Danish peat, are of recent species."

And further,—

" It has been stated that a stone implement was found [by Steenstrup] under a buried Scotch fir at a great depth in the peat."

Thus is it indicated that the fall of these early productions of Denmark must have been subsequent to the appearance of man in the locality.

At a higher level have been found remains of the oak, and at a still higher elevation remains of the beech.

" By collecting and studying a vast variety of such implements [of stone] and other articles of human workmanship, preserved in peat and in sand dunes on the coast, as also in certain shell-mounds of the aborigines presently to be described, the Danish and Swedish naturalists, MM. Nillson, Steenstrup, Forchhammer, Thomsen, Worsäe, and others, have succeeded in establishing a chronological succession of periods, which they have called the ages of stone, of bronze, and of iron, named from the materials which have each in their turn served for the fabrication of implements.

" The age of stone in Denmark coincided with the period the first of vegetation, or that of the Scotch fir, and in part, at least, with that of the second vegetation, or that of the oak. But a considerable portion of the oak period coincided with ' the age of bronze ' for swords and shields of that metal, now in the museum of Copenhagen, have been taken out of peat in which oaks abound. The age of iron corresponded more nearly with that of the beech tree." *

But we have further to determine, if practicable, the period of the time thus indicated ; and in doing this—or attempting it—some help may be derived from what are called by the Danes *Kjökkenmödding*, kitchen middens, or refuse heaps, which are found at certain points along the shores of nearly all the Danish islands. These are heaps of cockle, oyster, and other shells, mixed with bones of beast and bird and fish. Similar kitchen middens have been found elsewhere. Of those now under consideration it is remarked by Lyell that,—

* Marlot, *Bulletin*, cited p. 292.

"Scattered all through them are flint knives, hatchets, and other instruments of stone, horn, wood, and bone, with fragments of coarse pottery, mixed with charcoal and cinders, but never any implements of bronze, still less of iron."

These, then, must have been collected before the period of the growth of oaks on the borders of the peat bogs. The stone hatchets and knives had been sharpened by rubbing, and in this respect are one degree less rude than those of an older date, associated in France with the bones of extinct mammalia. The mounds vary in height from 3 to 10 feet, and in area some of them 1,000 feet long, and from 150 to 200 feet wide.

By various observations it has been attempted to determine the age of these mounds, and this is the question in which we, as students of forest science, are chiefly interested. Lyell follows up the statement of observations, by which it has been sought to determine that age, with the remarks,—

"No traces of grain of any sort have hitherto been discerned, nor any other indication that the ancient people had any knowledge of agriculture. The only vegetable remains in the mounds are burnt pieces of wood and some charred substance referred by Dr. Forchhammer to the *Zostera marina*, a sea plant which was perhaps used in the production of salt.

"What may be the antiquity of the earliest human remains preserved in the Danish peat cannot be estimated in centuries with any approach to accuracy. In the first place, in going back to the bronze age we already find ourselves beyond the reach of history, or even of tradition. In the time of the Romans the Danish isles were covered, as now, with magnificent beech forests. Nowhere in the world does this tree flourish more luxuriantly than in Denmark, and eighteen centuries seem to have done little or nothing towards modifying the character of the forest vegetation. Yet in the antecedent bronze period there were no beech trees, or at most but a few stragglers, the country being then covered with oak. In the age of stone, again, the Scotch fir prevailed, and already there were human inhabitants in those old pine forests. How many generations of each species of tree flourished in succession before the pine was supplanted by the oak, and the oak by the beech, can be but vaguely conjectured, but the minimum of time required for the formation of so much peat must, according to the estimate of Steenstrup and other good authorities, have amounted to at least 4,000 years; and there is nothing in the observed rate of the growth of peat opposed to the conclusion that the number of centuries may not have been four times as great, even though the signs of man's existence have not yet been traced down to the lowest or amorphous stratum. As to the 'shell-mounds,' they correspond in date to the older portion of the peaty record, or to the earliest part of the age of stone as known in Denmark."

Thus far we have advanced with firm footing under us; but here we are stayed, as if a voice had called and said, "Thus far canst thou go, but no further." Of what we have learned this is the sum.

In reference to the trunks of Scotch fir found in some of the bogs of Denmark, it is stated by Lyell,—

"This tree is not now, nor has ever been in historical times, a native of the Danish islands, and when introduced there it has never thriven; yet it was evidently indigenous in the human period, for Steenstrup has taken out with his own hands a flint instrument from below a buried trunk of one of these pines. It appears clear that the same Scotch fir was afterwards supplanted by the sessile variety of the common oak, of which many prostrate trunks occur in the peat at higher levels than the pine; and still higher the pedunculated variety of the same oak (*Quercus robur*, L.) occurs with the alder, birch (*Betula verrucosa*, Ehrh.), and hazel. The oak has now in its turn been almost superseded in Denmark by the common beech. Other trees, such as the white birch (*Betula alba*), characterize the lower part of the bogs, and disappear from the higher; while others again, like the aspen (*Populus tremula*), occur at all levels, and still flourish in Denmark."

From Vaupell's *Bögens Indvandring* we learn some curious results in regard to what may be called the natural succession of forest trees, which have been obtained in the examination of the bogs of Denmark by Steenstrup and himself. The bogs appear to have gone through some gradual process of desiccation; and the birch, which grows freely in very wet soils, appears to have contributed very effectually by its annual deposits to raise the surface above the water level, "and thus," says Marsh, "to prepare the ground for the oaks."

Between the extremes referred to as characteristic of the state of the ground in these peat bogs when the first and when the last of the trees borne by it appeared, there are innumerable gradations, and by the student of the habitats of different trees something may be learned in regard to what is the character of the soil in regard to humidity or aridity from the species produced by it.

In a volume entitled "Forest Moisture; or, the Effects of Forests in Humidity of Climate," * I have given details illustrative of the drying up of marshes on the growth of trees, and of the desiccating effects produced in marshes by forests in prolonged periods; and reference is here made to the indications supplied by successive crops of different kinds of trees of the progress of such operations.

From *Bogens Indvandring i de Danske Skove*, by Chr. Vaupell, we learn that within thirty years these bogs have yielded above a million of trunks of trees, and that they are found in depressions on the declivities of which they grew, lying with the top lowermost, always having fallen towards the bottom of the valley, and showing indications that they have fallen from age and not from wind, and that they are found in the order of superposition which has been stated.

According to Vaupell, the earlier forests of Denmark were composed of beeches, oaks, firs, aspens, willows, hazel, and maple, the first three being the leading species; but these have now been superseded by the

* Edinburgh: Oliver and Boyd, 1877.

beech, which at present is the tree which greatly predominates. But there are indications that it also may be supplanted. It is a tree more inclined to be exclusive than any other broad-leaved tree; but it is being encroached upon by the fir. This is attributed by Vaupell to the practice of removing the fallen leaves for bedding for cattle and other purposes.

"The leaves," says he, "belong to the soil. Without them it cannot preserve its fertility, and cannot furnish nutriment to the beech. The trees languish, produce seed incapable of germination and the spontaneous self-sowing, which is an indispensable element in the best systems of sylviculture, fails altogether in the bared and impoverished soil."

In connection with this he says,—

"The removal of the leaves is injurious to the forest, not only because it retards the growth of trees, but still more because it disqualifies the soil for the production of particular species. When the beech languishes, and the development of its leaves is less vigorous, and its arms less spreading, it becomes unable to resist the encroachments of the fir. The latter tree thrives in an inferior soil, and being no longer stifled by the thick foliage of the beech, it spreads gradually through the wood, while the beech retires before it and gradually perishes."

And thus Vaupell considers man is unconsciously, and to his own disadvantage, helping forward the encroachment of the Goths and Vandals on the nobler races at present in possession.

I have had occasion, in the volume cited, to say, "Thus may the succession of firs to beeches, if it should ultimately occur, be satisfactorily accounted for." And in a corresponding way may some of the previous successions of different kinds of trees be accounted for. I say not in the same way, but in some corresponding way; and in some corresponding way may much of the grass and herbage, which previously covered the forest ground, have given way to the forest trees, not because the conditions were absolutely unfavourable to them, but because being also favourable to the growth of trees, and perhaps comparatively more so, or otherwise simply from their more sturdy constitution, these, having gained a footing, maintained it, and the others were overpowered.

It is interesting, and in some cases advantageous, to read the pre-historic records preserved in nature. And an acquaintance with the natural history of trees found in the Danish bogs may throw light on the gradual progress of the desiccation of these bogs, and on the contemporaneous circumstances in which they existed at successive periods of their history.

II.—NORWAY.

I ENTERED Norway at Christiansand, and proceeded thence by Norwegian steamer to Christiania. I arrived about 6 a.m., and as the steamer, of which there are one or more sailing every day, had left three hours before, I had the day at my disposal, and walked first to the cemetery, and after breakfast returning thither I rambled about on the lovers' walk, extending over miles of well-kept footpaths, winding about interminably upon rocks and rising ground to the left of the land as you approach by sea; and I found some of the views of the Fiord, and of adjacent fiords, dotted with islands, exceedingly beautiful. After dinner I took a sail by smaller steamer up the Torristal river, returning by seven p.m. This little voyage, a similar one up the Topptal river, and one which would have taken me out in the fiords, were all recommended to me. My choice was made capriciously. I believe, from what I was told, that I would have found either of the others equally pleasant, but different. I found my morning ramble and my afternoon trip together a good preparation for visiting more extensively the beautiful and romantic scenery of Norway. On my return voyage on the Torristal river I met with a timber merchant, who had that day completed the purchase of a ship-load of wood which was still growing in the forest, and who courteously and frankly informed me in regard to the transaction. The farmer, a small landed proprietor, had engaged, on the terms agreed to, to fell, prune, and deliver in the river the quantity of wood required in logs of a specified average length and girth. All the *débris* remained his property, and might be cut up for firewood and used or sold, or otherwise disposed of as he might please. And the logs would be, when delivered and accepted, floated down to Christiansand and shipped by the purchaser.

I have used the expression farmer or small landed proprietor, for such is the character of most of the inhabitants of this district. The banks of the Topptal river are dotted with houses and mansions, which speak of greater wealth or more extensive possessions; and many of the farmers are dependent on the cutting and sale of wood to enable them to pay their rent. The logs I found to be of but limited dimensions, and I asked the timber merchant if the exploitation adopted did

B

not tend to devastate the woods. "Look," said he in reply, "there, are young trees growing in every nook and corner down to the water's edge on both sides of the river, throughout the whole course of our voyage. And so is it for miles inland. As fast as we fell others grow." I called his attention to the small size of the logs, and told him of what was being done in Sweden and elsewhere for the conservation and improvement of forests ; but he only laughed, as if that were a thing altogether unnecessary here, and one which it would be ridiculous to propose, and he called my attention to the floats of timber we were constantly passing.

At one mill at Christiansand it is said 70,000 trees are thus floated down and sawn up every year, and there are several other saw-mills in the town. At Vigelund, about ten miles above the town on the Torristal river, a little way above the rapids of the river to which the steamer goes, there is another saw-mill. At the fall on which this mill stands may be seen what I shall hereafter have occasion to refer to as characteristic of the water transport of timber trees in Norway. The author of a volume entitled "Frost and Fire," describing the passage of this fall by trees, says, "'At every moment some new arrival comes sailing down the rapids, pitches over the fall, and dives into a foaming ground pool, where hundreds of other logs are revolving and whirling about each other in creamy froth. The new comer first takes a header, and dives into some unknown depth, but presently he shoots up in the midst of the pool, rolls over and over, and shakes himself till he finds his ,level, and then he joins the dance. There is first a slow sober glissade eastward across the stream to a rock which bears the mark of many a hard blow. There is a shuffle, a concussion and a retreat, followed by a pirouette sunwise, and a sidelong sweep northwards up stream towards the fall. Then comes a vehement whirling over and over, or if a tree gets his head under the fall, there is a somersault, like a performance in the Halling dance. That is followed by a rush sideways and westward, when there is a long fit of setting to partners under the lee of a big rock ; then comes a simultaneous rush southwards, towards the rapid which leads to the sea, and some logs escape and depart, but the rest appear to be seized with some freak, and away they all slide eastwards again across the stream to have another bout with the old battered pudding-stone rock below the sawmill : and so for hours and days logs whirl one way, in this case against the sun, below the fall, and they dash against the rounded walls of the pool. Such is the effect of these concussions that above the fall it has been found necessary to protect the rock against floating bodies so as to preserve the way of the stream. It threatened to alter its course and leave the mill dry, for the rock was wearing rapidly. Lower down, nearer the sea, is a long flat marsh, between high, rounded cliffs; and there these mountaineers, floating on to be sawn up, form themselves into a solemn funeral procession which extends for miles; and it may be noticed that the course of this stream of floats is always longer than the course of the river's bed; for the water is slowly swinging from side to side as it flows, and the floats show the course of the stream and its whirling eddies."

The Halling dance referred to is a Norwegian dance, reminding one at once of the sailors' hornpipe and the Highland fling, but still more vigourous and exhausting to the dancer, being diversified with somersaults such as are here alluded to.

On my return to Christiansand in the evening I at once took possession of my berth on board the steamer, which, like others in which I have sailed on these northern seas, was more luxurious in its arrangements than are those in many of the sea-going steamers on the British coast.

We sailed at three o'clock in the morning, and from that time till near nine o'clock in the evening, with the exception of about an hour and a half in the middle of the day when we left the coast, we were the whole way sailing onwards among islands, rounding them and passing them as if on a pleasure trip—there being always rocks to seaward to break the roll of the waves, and secure for us placid waters,—and looking in on every village on the coast, while breakfast, dinner, and supper were served with all the northern whets and appetisers of which many a one has heard.

Steamers from Britain to Christiania generally avoid the coast, so that this sight is lost until they enter the Fiord. Such as is the Fiord, such was the whole course, with the trifling exception of the hour and a half spoken of. I was reminded of a voyage through the Thousand Isles of Lake Ontario; but the scene was different. Here the islands are rocks, but not rocks rough and rugged—rocks of granite planed down and smoothed by glacial action, more like clean and white and sparkling banks of mud than are rocks on a sea-girt shore. It required no effort and but little fancy to picture them as an ocean bed rising above the sea, when, according to the Hebrew cosmogony, God said, "Let the waters under the heavens be gathered together, and let the dry land appear." And again, "Let the earth bring forth grass, the herb yielding seed, and the fruit tree yielding fruit after his kind: and it was so."

There, were the bare, rounded granite rocks, without a blade of vegetation; there, were others with only a lichen, or a moss or a grassy, or flowery green spot. The former was on the dry rock, the latter on any crack or hollowed basin; and there, where there was a wider rent or a cup-like basin containing a handful of earth, a sapling tree ; and there, an island not much larger, covered with trees to the water's edge, and there, larger islands, or the main land, with high rising hills clothed with wood and forests beyond. Again and again I felt that day as if I were alone with God, or rather with His work, as His work is described by Wisdom in the Book of Proverbs: "When there were no depths, I was brought forth ; when there were no fountains abounding with water. Before the mountains were settled, before the hills was I brought forth : while as yet He had not made the earth, nor the fields, nor the highest part of the dust of the world. When He prepared the heavens : when He set a compass upon the face of the depth : when He established the clouds above : when He strengthened the fountains of the deep : when He gave to the sea His decree, that the waters would not pass His commandment: when He appointed the foundations of the earth."

The effect was heightened by the general absence of animal life, excepting at the towns and villages. Once or twice a cow was seen, once or twice a bird on the wing, and once a realization of Kingsley's picture of the sea-gull on the All-alone stone far out at sea ! This was as we left the islands shortly after noon. There were four gulls struggling to maintain their footing on a little projecting rock far out at sea, washed over by a wave produced by our passing vessel. And here and there a solitary house, neat, painted, and clean, might be seen, or an island where road there was none but the highway of the sea, as if man were only beginning to appear upon the earth.

From Christiania I took a trip towards Drammen, and saw what Norway is under cultivation ; and my journey towards Aarnaes and Charlottenburg gave me, in the first part of that journey, an opportunity of seeing under cloud and rain, Norway in a condition similar to that of our moorland districts in Britain. As if the former trip had shown what the earth was when the earth was young—inhabited by man and cultivated, but young—this seemed to show what appearance the earth puts on in old age; and in journeying from Aarnaes towards Charlottenburg I found yet another aspect presented.

There are extensive districts in the vicinity of Glasgow, of Newcastle, and of Durham, where it appears to be coal, coal, coal, and iron and coal, but chiefly coal, which constitutes the one article of transport there and product of the locality.

In America, again, in travelling through the so-called oil district lying between Pittsburg and the eastern shores of Lake Ontario, it is oil, oil, oil,—oil everywhere,—what seem interminable trains of waggons, but oil cisterns all of them, and pipes like large water-pipes or drain pipes—all conveying oil.

Here it is wood, wood, wood, or perhaps I should say timber, timber, timber and wood everywhere. Wood by the roadside, trucks laden with wood, wood piled at the stations and on the fields, and last of all a river covered with wood and floating timber. This is the Glommen, here a broad river, and apparently deep ; near the railway bridge by which it is crossed, the logs have been collected into floating islands of wood, begirt and confined by a chain, of which the links are logs, logs with a hole bored at either end, and tied one to another by withes. As we proceed we see the river bearing hundreds and thousands of logs onward to this gathering-place. The size of the river compared with the size of these suggests the idea of some boys having emptied into a brook a hundred, or a thousand, or a hundred thousand boxes of matches, and we looking on seeing them floating away. Again and again we came upon a little fall, one of three or four feet, and there they came tumbling down sometimes sideways, sometimes slanting, sometimes head foremost, kicking up their heels in the air.

The river is broad, it comes curving along through woodlands, these partly concealed, and I felt as if I could realize the graphic picture given by Hugh Miller of a river in pre-Adamic times bringing down the forestal products which afterwards were converted into fields of coal.

The Glommen is the principal river in Norway. It originates in the lake Oresund, under the 62° of north latitude, and runs southward about 90 miles through a rugged channel full of cataracts and shoals. One of its confluents is the Worm, which flows through Lake Mios. Before their confluence it is as large as the Thames at Putney, and about 20 miles below this it flows into the sea at Frederickstadt. Its highest cataract is that of Sarpen, which is 60 feet perpendicularly, and is not far from its influx into the sea.

In travelling thus far one meets chiefly with a stalwart race of yeomen, presenting very much the same general appearance as do Americans in rural districts in the United States, or as do sub-stantial Dutch boors in the inland districts of the colony of the Cape of Good Hope. But in Christiania there is a museum of Scandinavian curiosities, amongst which are life-size figures of Norwegian peasants in picturesque national costumes, which I had previously seen do good service at one of the International Exhibitions, either that at Paris in 1867, or that at Vienna in 1873—and which have been secured for permanent exhibition here. I may mention in passing that I was struck with the resemblance of many of the Norwegians of all classes, both men and women, to personal friends of my own in Scotland.. I have named a dozen, and might have named a score of friends whose figure, gait, and countenance I found completely reproduced; and I have never found this in other lands.

The author of the volume entitled "Frost and Fire," already cited, noted like similarities between words in the Norwegian and the Scottish languages. At one place he writes, "A 'cow' here is ' nout,' as in Scotland; to walk is ' tootla,' ' toddle.' The Saetar girl and the guide who lived about seven miles apart, have different dialects, at least so they said; and in this they are like the natives of all mountainous districts, from the Yorkshire dales to the antipodes." And again, a little later, " We rowed over the fjord to the place where the priest sleeps when he comes to preach; there was not a living soul about the place when we arrived, so I got in through a window and took possession of the priest's room. As it grew dark, people came tumbling in from the woods where they had been working, and we had a party round the fire at one of the houses. I could not understand half they said, for I had now got into a fresh dialect; but I fancied my hostess was a witch or a doctress, for men purchased mysterious oil from a bottle, which was carefully weighed, and one pretty girl had a long conversation about some one who had been sick, and who was now " frisk." Presently the door opened, and the husband, with a wet bag and a creel of live fish, tumbled in, and then we all sat with our faces lighted up by the wood fire, chattering like a flock of gulls; while a little girl, who woke up at the noise, kept screaming like a young cormorant from its nest, ' Moor, gie me fisk;' that is, ' Mother, give me fish.' " I have met with the same thing in Finland, which was formerly connected with Sweden, and is geographically connected with both Norway and Sweden by Lapland. There I have met with *a cruse o oely*, a cruse of oil, and *flitting tyd*, or the flitting time.

By another route, already indicated, the traveller may pass more or less rapidly through the greater part of Denmark, and cross from Frederickshavn to Gottenburg, in Sweden, by steamer sailing in connection with a railway train, and arriving at Gottenburg in the evening, about eight o'clock; or Norway may be reached by steamers from Hull to Gottenburg, and to Christiania; or by steamer from Leith to Christiansand; or by steamer from Hull to Bergen, touching at Stavanger; and there are steamers from London to one and another of these ports.

In Norway, in the interior, the traveller may find that he has entered a land, the most marked characteristics of which are forests and fiords; "forests whose vastness and shade, and solitude and silence, banish in an instant from the mind all associations with songs of birds and gay sylvan scene;" and combined with these are also "lakes whose deep seclusion put to flight images of mere grace and beauty,—valleys, which from their depth and gloom one might fancy to be the avenue to abodes of more mysterious creation; mountains, whose dim and rugged and gigantic forms seem like the images of a world which one might dream of but never behold."

In such terms is the country spoken of by an intrepid traveller, writing under the *nom de plume* of Derwent Conway; and in similar terms is it spoken of by others. By one traveller, Norway is spoken of as a land "whose only charm is to be found in her dim mountains, her silent forests, and her lonely lakes."

Another, Edward Price, an artist, who traversed the land and looked upon every scene with an artist's eye, speaks of Norway as a country "which surpasses every country of Europe in the depths of its fiords, and in the grandeur of its forests and forest scenery." Having landed at a distant point, and traversed the land, chiefly on foot, seeing thus much which could not otherwise have been seen in the course of his tour, he reached the capital; and of what he saw as he approached it he thus writes: "Luxuriant pasturage and crops, giving rich promise of an abundant harvest, lay on every side. Wood was no longer the great staple of the land, but was scattered over a charming undulating country only in such quantity as served to shelter the fields and beautify the landscape; nor was it now confined to fir, but included all the variety of trees which we are accustomed to find in the temperate latitudes. The Christiania fiord, spotted with its islands, and seemingly environed with its finely wooded banks, formed innumerable bays and creeks, all calm and pellucid beneath the warm rays of the noonday sun." Thus did it appear when approached from the land; and in accordance with this is the account of its appearance as approached from the sea.

With an account of this we are supplied by the Rev. Henry Newland, in a popular little work entitled, "Forest Life in Norway and Sweden, being Extracts from the Journal of a Fisherman"; and throughout this little work are dispersed several graphic sketches of woodland scenery, as viewed with the eye of a sportsman.

Derwent Conway's tour was made, if I mistake not, in 1827. Price travelled through Norway in the year before this; but fifty years have made no great change in the general features of the country. It was in 1833 that I first went to the north of Europe. In that year Barrow devoted a summer vacation to a tour through Norway, having previously made short excursions through Russia, Sweden, and Denmark; and not a few of his sketches will be found in keeping with those given by the other tourists cited.

Somewhat nearer to the present time was published in 1850 "Rambles in Norway," by Thos. Forrester, Esq., in which there are interspersed in the narrative many little sketches of woodland scenery, and of the woodland population.

From any one of these publications some definite idea of the physical geography and woodland scenery of Norway may be obtained.

From conversations with Mr. Charles S. Inglis, of Edinburgh a gentleman who travelled extensively in the country, roughing it, and rambling in regions and districts out of the ordinary route of tourists, and who has published his observations under a *nom de plume*, I first learned that in the northern portion of Norway the land presents the appearance of table-lands, or comparatively level plateaus, cut up by what may be called ravines rather than valleys, which are sometimes more than a thousand feet in depth ; these can only be crossed by zigzag tracks or roads, descending the precipitous declivity on one side, crossing a streamlet at the bottom, and ascending in a similar zigzag way a corresponding precipitous ascent on the other ; and in other places, that of isolated hills and mountains, scattered about in what looks like studied confusion, sometimes standing apart and alone, but as frequently in groups of more or less irregularity, and of greater or less extent, and sometimes, but that rarely, taking a form not unlike a mountain range. Towards the south the country assumes gradually a more level aspect, but it does so without losing altogether its hilly character. The result of the whole is that about two-thirds of the country is at an elevation of upwards of 2,000 feet above the level of the sea, which is considerably above the range of forests trees in that land.

The forests in Norway are by no means so extensive as the frequent mention made in England of Norway timber may lead those who have never visited the country to imagine. They are generally found along river courses. They extend from half a mile to three or four miles from the banks of the river, and up the precipitous hill-sides beyond. Sometimes the continuity is broken abruptly on the river-bed by perpendicular cliffs ; but the forest extends on the table-land above, like a dislocated geological stratum, or the further side of a dyke or fault.

Many of the forests are private property ; others belong to commercial proprietors. In both classes of forests the right to fell timber is generally let to contractors possessed of large capital, by whom arrangements for felling wood upon an extensive scale are made.

Previous to the introduction, of late years, of an improved forest economy, the system of exploitation or working usually adopted was one intermediate between that known in France as *Jardinage*, felling only such trees as were desired, and that known as *à tire et aire*, in which the forest is divided into as many sections as periods required for the reproduction of the crops, and these are cleared in succession, but only one in each period : the *coupés*, or fellings in different periods in these Norwegian forests not being regulated in extent by precise measurement, but being determined by the convenience of the contractor ; and only trees suitable for his purpose being felled. These are generally trees, on an average, a little under two feet in diameter ; and all such are felled, leaving after them but a poor and scraggy crop of growing trees to replace, in course of time, if they can, what has been removed.

In Norway there is no lack of means for transporting the felled timber by water to the coast. In many places the felled trees, stripped only of their larger boughs, are tumbled into a mountain stream, to be by it borne to the nearest river or lake; in others, they are shot along artificially constructed slides, leading to some lake or river. These slides are in structure intermediate between the *chemins à trainaux* and the *lançoirs* or *glissoires artificiels*, used in France. They are about 5 feet wide. Sleepers are laid across the line at about equal distances apart, and upon these are laid, lengthwise, trunks of young trees about 5 or 6 inches apart, and often so arranged that those at the sides are somewhat higher than those in the middle to form a groove of sufficient depth to keep the shot timber in the slide. In some cases these slides run directly down the declivity to the river or lake to which they are destined to convey the timber. In other cases, they run across the side of the hill in a slanting direction. In some places earth is removed to allow of the desired angle of inclination being secured. More frequently this is attained by the slide being supported at places by piles of earth or beams. When necessary, they are carried on supports across small valleys, or watercourses, separating the forest on the one side of a mountain from the forest on the side of another ; and occasionally there may be seen their straight course altered by an angle more or less abrupt. At such places there is generally raised at the outer angle of the slide a bank against which the trees may strike in their descent and then recoil into the new direction : these by the new direction thus given to their course, go on, sometimes head foremost, and sometimes making first a complete somersault or revolution. In general also a workman, or it may be two, or even three, are stationed at these points with long poles to aid at the time the movement of any trees which might otherwise be in danger of sticking fast and blocking the way. The slides in general lead to pools of considerable depth in a lake or river. Into these the trees ofttimes descend more than their entire length, starting up again vertically before setting off anew on their course.

Much *débris* is found all about such spots; but it is comparatively seldom that logs are seriously damaged. The quantity of splinters may be, to a great extent, composed of the lesser branches left on the tree when placed in the slide.

Brands or marks may be seen upon some logs. These are, I presume, the marks of the woodcutter or the contractor, made to enable each one to claim his own property should logs belonging to different proprietors get mixed together.

The logs are carried down by the river, and if the river fall into a lake, they are—at the embouchure of the river—collected and formed into a raft ; and such rafts are sometimes towed by a little steamer across the lake to its outlet. If the stream flowing thence be smooth, they may be floated further as a raft; if it proceed over waterfalls in its course, the logs are unchained and allowed to float down apart, to be reformed into a raft below these, if circumstances allow of this. Notwithstanding the care which may be taken, many logs are stranded on the banks of the lakes and rivers. The logs are cut into size and shape for the foreign market by saw-mills near the coast, which are driven by water-power.

Saw-mills of the simplest structure, consisting only of a water-wheel and a circular saw, fixed apparently on the same axle, are common appendages to farmhouses in the country. They are employed in cutting up the firewood required for the family and their retainers.

By Forrester it is mentioned that in some cases two years have been occupied in the transport to the sea of timber cut in the upper mountains. From this some idea may be formed of the difficulties which have to be overcome, and which are overcome, by the indomitable industry of the people.

The forests consist almost entirely of the Norway spruce fir (*Abies communis*); and in some parts of the country tar is manufactured.

III.—SWEDEN.

As the union of Scotland and England has had the effect of leading most foreigners to think of these countries as one, with a general but somewhat vague idea, which is substantially correct, that the one land is the northern and the other the southern half of the island, so the union of Norway and Sweden has had the effect of leading many of our countrymen to think of these countries as one, with a general if somewhat vague idea, which is also substantially correct, that the one is the western and the other the eastern half of the peninsula.

In so far as forests and forest lands are concerned it may be convenient to treat the two countries as one; but when the question to be discussed relates to forest economy this can no longer be done; and if the question arise, Whence comes the difference here? we are brought to the discovery of the fact, if it was not previously known, that Norway and Sweden are inhabited by two different nations, having different laws, different governments, united only in having the same king, so that the union is more like that which formerly subsisted between Great Britain and Hanover than that subsisting between England and Scotland, where, though there be different laws in force in the different countries, there is but one government common to both.

Sweden may be reached by any of the routes detailed as practicable for reaching Norway. From Copenhagen the traveller may go to Stockholm *viâ* Malmo by railway in a day and a half The journey from Gottenburg may be made by railway in a day, or by steamer and the Gotha Canal in two days, the most interesting portion of the scenery being passed by daylight, and this route taking the traveller along by the famous Falls of Trollhättan. Should the route by Christiania be preferred, the sail up the fiord will be greatly enjoyed. From Christiania pleasant trips may be made to the Miosen Lake and other places; but the mountainous region of the peninsula lies further to the west, and stretches away far within the Arctic Circle.

To the traveller desirous only of visiting Sweden it is preferable that he should proceed from Gottenburg to Stockholm by the Gotha Canal, or return by that route if he go to the capital by some other course. It will lead him through scenery grand, magnificent, and beautiful.

My route leading me through Christiania, I entered Sweden by Charlottenburg, which, like some of the other places at which I rested, or through which I passed, I found to be pleasantly situated and cheerful. The journey made continuously by rail occupies about eighteen hours. On the way I saw nothing like the mountain scenery of Norway, but several extensive lakes in a level country, apparently only some three or four feet higher than their surface; and I saw nothing of the forests of Sweden, which are situated in part somewhat to the south and in far greater part further to the north.

The forests of Sweden are extensive, covering about two-sevenths of its entire area. The varieties of timber, however, are few. In the north the pine, birch, and fir are the principal trees; in the central parts the ash, alder, willow, and maple are also common; and in the south the oak, beech, elm, and lime are met with. In the plain of Scania the mulberry, chestnut, pear, apple, and walnut trees flourish.

In Sweden, says Marny, a French writer on the forests of Europe, the woods are numerous, but little productive, and we only rarely meet with vast forests. Dalecarlia, Wermeland, and the district of Orvebro are the only central counties where arborescent vegetation attains sufficient energy to cover with wood a large extent of country. There the Coniferæ almost always constitute the basis of the forests. Sometimes, however, the birch replaces them, notably in Oster-Goth-land. Sweden, like America and Siberia, has her forest fires, which deprive in a brief time a whole forest of its shade; and vegetable life once destroyed revives only with difficulty on this frozen soil.

In Norway the forests are more extensive; they stretch along the Scandinavian Alps, which separate this country from Sweden. The birch there reaches an altitude of 1,200 feet above the sea.

In the diocese of Bergen the fir has still the gigantic proportions seen in the forests of Switzerland and Germany, but more to the north its size is diminished to stunted proportions, and at the polar circle it has totally disappeared; whilst in Swedish Lapland it advances yet to two degrees beyond it.

In Norway it is the birch which really serves as a ladder to vegeta-tion. It is the measure of its energy, and it marks by the different states through which it passes, in proportion as it rises in altitude, the degree of weakness of vegetable life. To the weeping birch succeeds the *Betula acer*, which replaces the white birch; after which comes the birch of the prairies, which passes in its turn through different gradations of size, and which at the polar circle is nothing more than a stunted shrub of pyramidal form, and covered with moss.

In Sweden of late years strenuous and successful efforts have been made to introduce into the management of the forests the latest improvements in forest management adopted in Germany and France, and to regulate the national forest economy in accordance with the most advanced forest science of the day. In regard to the measures adopted, and the history of the movement from its commencement to the present time, detailed information is forthcoming whenever it may be desired and occasion serve.

In a pamphlet entitled "The Schools of Forestry in Europe: a Plea for the Creation of a School of Forestry in connection with the Arboretum at Edinburgh." * I have given a translation of the regulations of the Forest Institute and Primary Forest Schools of Sweden, which were sanctioned by the king under date of May 25, 1860.

The following is a translation of the existing later regulations :—

HIS MAJESTY'S NEW ORDINANCES FOR THE PUBLIC SCHOOLS OF FORESTRY IN THE KINGDOM.

Stockholm Palace, 15th September, 1871.

Chapter I.—The Forest Institute.

Section 1st.—The Forest Institute, which stands under the inspection of the Forest Department, and is presided over by a director, has for its end to educate able forest managers by free instruction.

Section 2nd.—At the Forest Institute in the Royal Zoological Gardens at Stockholm, a convenient locality is open at all times, containing partly lecture halls and the necessary buildings, also space for the collections necessary for instruction, partly also a portion of ground for nurseries, and the planting of trees, and exercise grounds.

Section 3rd.—In the Institute instruction shall be given in the partition of forests, cultivation of forests, forest technology, the doctrine of climate and soil, forest botany, forest entomology, forest zoology, hunting, mathematics, the institutes of forestry and hunting, book-keeping and transaction of business, map-drawing, weighing, and the art of shooting.

Section 4th.—For the practical yearly exercises appertaining to the instruction, suitable forests must be appropriated by the Forest Department, in case the fields adjoining to the Institute are not sufficient.

Section 5th.—The course of instruction, which begins with the month of June each year, and for which pupils may be taken so long as there is room in the Institute and means for their support, continues two years, with a corresponding division of the subjects of instruction for each year, so that the teaching in the Institute, together with the practical exercises in its neighbourhood, goes on from the beginning of October to the end of May, with three weeks' vacation at Christmas ; and the summer months are to be applied to forest measurement, together with other practical exercises, in the woods appropriated by the Forest Department for this purpose.

Section 6th.—With the division of the subjects of instruction laid down as follows, the instruction shall be carried out by the director as forest lector, and three lectors, who shall constitute the college of the Institute. On the invita-

* Edinburgh: Oliver and Boyd, 1877.

tion of the director, the body of teachers shall meet together for common consultation on matters affecting the work of instruction as may be demanded. For assistance in carrying out the practical exercises in the division of forests, the Forest Department may, on the director's proposal, instal two tutors (*repetitorer*) for the time of those exercises, or one, as may be found necessary.

Section 7th.—For the support of suitable pupils who distinguished themselves through diligence, ability, and good cultivation, also of graduated pupils who may be employed as tutors, stipends are to be supplied to the amount which may be fixed upon. The Forest Department shall appoint candidates to these stipends on the recommendation of the college of teachers.

Section 8th.—The director in the Forest Institute must have a thorough insight into all the subjects of instruction taught in the Institute, and a complete theoretical as well as practical knowledge of forest economy. His duties are—

1st. To carry out with attention the course of instruction, and take care that this, without setting aside theory, take a practical direction, and also to see that secondary subjects do not develop themselves at the expense of the instruction in forest economy. The director must send in in the months of September and January to the Forest Department for examination and confirmation the papers of the college of teachers, in acccordance with the above-named plan of the order of work for the session following.

2nd. Carefully to preserve good order, good moral conduct, and diligence among the pupils of the Institution, and with this view to carry out the arrangements laid down.

3rd. To send in to the Forest Administration before the beginning of each quarter the estimates for the stipends of the candidates.

4th. To send in yearly to the Forest Administration before the month of April a report relative to the action of the Institute during the previous year, and also to give in to the Forest Administration his opinion regarding the various questions concerning the Institute which may be raised by himself or the college of teachers; also to mention such business belonging to the Institute as does not rest upon himself immediately.

5th. To watch over the preservation and safe keeping of the buildings appropriated to the Institution, together with the inventories and collections of the Institution, of which the director has to keep a complete list.

6th. To receive every quarter, of the Forest Department on requisition by his Imperial Majesty and the kingdom's Chancery, the money set aside for the payment of salaries and the support of the Institute, together with the other moneys which shall be made use of according to the given prescriptions for the payment of each year's charges, the papers belonging to which, together with an inventory of the property of the Institute, shall be handed in in the month of February following to the Forest Department.

7th. As to the forest lector, conformably to the precisely laid down order of work, he shall instruct the pupils in the division and the culture of forests and also in forest technology as an introduction to the already mentioned branches of knowledge, especially in that of forest statistics and history, together with the literature connected with forest science, with reference to the final examination of the pupils in the said branches of knowledge;

it belongs to the director but to conduct the practical exercises which belong to his subjects of instruction, and also to prepare the pupils to visit occasionally the wharves, sawing establishments, and other places in the metropolis or its neighbourhood where forest products are made use of and manufactured : in which visits the director must accompany the pupils, and communicate on the spot the information suggested by the things seen.

8th. To superintend the management of the woods which are put by the Forest Department for the instruction of the pupils under the guardianship of the Institute. The director shall live in the buildings connected with the Institute.

Section 9th. The lectors in the Forest Institute shall each in that branch of science which is entrusted to him possess thorough and comprehensive knowledge, and both communicate instruction in the subjects taught according to the fixed order of instruction, and arrange a final examination in the branches before dismissal of the pupils.

Section 10th. The lector in the branch for natural science has to give instruction in the branches of science relating to climate and soil, so far as applies to forest economy; in forest botany, with vegetable physiology and vegetable geography ; the branch of science relating to insects profitable or harmful to woods, with the methods of exterminating the latter; the branch of science relating to Scandinavian wild animals and birds, with respect to the preservation of forests and hunting; also the art of hunting; and to exercise practically the students in the various branches of natural history by visiting with them the museum of the Academy of Sciences, and by natural history excursions with the pupils through the grounds of the Institute, or in the neighbourhood of the metropolis; and also to promote the student's proficiency in the art of shooting in the shooting gallery attached to the Institute.

Section 11th. The lector in the mathematical sciences has to instruct in mathematics applied to forest economy; in the art of building, with respect to the buildings for roads, waters, and other necessary purposes occurring in forest technology ; the elements of land measuring, surveying, and map-drawing ; and also to exercise the pupils practically in the grounds of the Institute in taking levels, so as to promote their proficiency in diagram drawing.

Section 12th.—The lector in jurisprudence has to instruct in the legislation of forest economy and the game laws, together with the methods of carrying on legal proceedings and the management of forest service and book-keeping.

Section 13th.—The tutor who on the ground of proved skill in the division of forests is installed for the assistance of the director in the practical exercises in this branch has, with respect to his official duty, to place himself at the director's disposal in the discharge of the duties of his office.

Section 14th.—In order to obtain admittance as a pupil in the Forest Institute, applications must be lodged with the director before the middle of May, and the applications must be accompanied by certificates that the applicant is not less than eighteen and not more than twenty-eight years of age, that he possesses a good character, that he has a sound and healthy constitution, is not affected by any incurable disease, and that he has been declared, on the occasion of his final examination on dismissal from the higher school, fit to go to the university. The applicant ought, besides his certificate of dismissal from the classical branches, to present a special certificate from a proper person that he has the necessary knowledge in mathematics, physics, and chemistry, and also that he has either gone through a forest school, or that he has, at least, during one year, obtained practical acquaintance with forest management and forest measuring by some official on the forest establishment properly appointed by the Forest Bureau.

This certificate ought to contain some information as to the whole of the forest work in which the candidate took part, together with proof under his own hand that he can measure a field by means of the tables for land-measuring.

The said certificate must be examined by the college of teachers, and thereon, if they shall see fit, the candidate shall be admitted as a pupil into the Forest Institute.

In order that the Forest Bureau may be in a position to state how far exception may be allowed from the above-mentioned rule, the applicant must also state his age.

Section 15th.—It is incumbent on pupils in the Forest Institute to lead a correct moral life, to treat superiors with consideration, and obey their orders, to wait on their instructions with attention and diligence, to carefully execute the work which they receive to do, to preserve carefully such things belonging to the Institute as may be entrusted to them, and they must repay whatever they damage; and in all things strictly conform to the laws of the Institute.

Section 16th.—In case of the pupil not fulfilling what, according to the preceding sections, is incumbent upon him, and taking no heed of the warnings given him by the director, such pupil may, after conference with the other teachers, be expelled.

Section 17th.—In order to receive a proper certificate of dismission the pupil must manifest in the public examination a well-marked acquaintance and ability in the collective subjects of instruction mentioned in section 3rd, and must have prepared a complete plan for the division of forests as well as for the cutting of roads and the thinning of trees, and also for the laws of forest cultivation.

The certificate of dismission must be made out by the college of teachers, according to the formula laid down by the Forest Bureau, and be handed to the pupil by the director.

Section 18th.—The plantation keeper in the Institute who has been appointed to reside there shall place himself at the disposal of the director for carrying out the instruction appointed by the Forest Bureau.

Chapter II.—District Forest Schools.

Section 19th.—The aim of these forest schools is through gratis instruction to form good foresters. These schools stand under the oversight of the Forest Bureau, and each of them under the visitation of the forest inspector in whose district of service the school is situated; and each forest school be presided over by a president, who is at the same time the teacher of the school, but for his assistance in the instruction and the corresponding practical exercises a forest overseer is to be appointed.

Section 20th.—The estates proposed for these forest schools, in the public forests, in different parts of the country, shall be furnished with lecture halls and the needful houses for the president and the forest overseer, as also with lodgings for a certain number of pupils.

Section 21st.—Instruction in the forest schools shall be communicated in the following subjects:—Writing and the formation of a good running hand, correct spelling as exercised by writing to dictation, arithmetic as far as the rule of three and decimal fractions, together with a knowledge of the Swedish system of money, weights, and measures;

Palisading, together with measuring, map-drawing, and calculation of his own figures, according to different scales, together with the calculation of the contents of the solid figures occurring in forest economy;

The natural history of the forest trees of Sweden, and their requirements with respect to soil, situation, &c.;

The natural history of the most frequently occurring animals, noxious or profitable to forests, and also the most important species of animals as respects hunting;

The general elements of forest economy to the extent that is required to make plans for forest management, and to carry out those plans needful for forest cultivation, to discharge the duties incumbent upon a forest servant, and to decide upon the most suitable application of wood,—all which ought chiefly to be taught by practical exercises and labours;

The sawing (or cutting up) of timber, the burning of charcoal, and also tar extraction and potash burning; the regulations for hunting and forests, so far as is demanded on the part of a forester; the making up of such reports, lists of day's-works, and so forth, as have to be given by a forester, together with an acquaintance with the hunting of wild animals, and the art of shooting.

Section 22nd.—The course begins every year on the 1st October, when pupils, according to the determination at the entrance examination, which takes place in September, will be received in such numbers as circumstances permit, and will undergo instruction, with the exception of a single week at Christmas, until the following 15th June, during which time the pupils will go through the course of study detailed in the preceding sections, together with the corresponding practical exercises.

Nevertheless the Forest Department may, if the practical exercises in consequence of climatic or other conditions of the locality where the school is situated require an alteration in the above-stated arrangement of time, arrange the same otherwise, provided that the time of instruction be not shortened.

Section 23rd.—For a certain number of pupils the proposed means of support to a separately determined amount shall be allotted, according to the president's examination, to the pupils who show themselves the most deserving of this, and these shall likewise obtain lodging in the forest school.

Section 24th.—After the conclusion of the course of instruction a public dismissal examination in presence of the proper Inspector of Forests shall be appointed for the pupils, when those who show themselves capable of undertaking the duty of foresters shall obtain a certificate of dismissal.

Section 25th.—The president of the forest school must have gone through a full dismissal examination, and he afterwards must on his own responsibility superintend the district.

It is incumbent upon him—1st. To be responsible for the proper conduct of the school, the preservation of good order, the morality and diligence of the pupils ;and at the same time he shall be warden of the school buildings and its other belongings.

2nd. That in all the branches of instruction laid down in section 21st he shall conduct the instruction, and specially see to it that the pupils are exercised practically for the purpose of obtaining sufficient knowledge and ability as foresters.

3rd. That he shall draw quarterly from the person who has it in charge from the king's majesty, and dispose properly of the money set apart for the

C

school; also that he shall submit an account in each year to the Forest Administration, in the course of the following February one as to the application of the money that has been allotted, and in the course of the following April a yearly report as to the work of the school.

4th. To appoint an entrance examination for those seeking admission as pupils, and to receive the pupils on the ground of the same; also immediately after the close of the course to conduct a dismissal examination, and thereupon to hand to the deserving pupils a certificate of dismissal according to the formula provided by the Forest Administration.

Section 26th.—The Forest Overseer must have obtained a certificate of dismissal from the forest school, must show dexterity in the carrying out of forest works, must be able to write, and must be well reputed for steadiness and good order; he shall be under the order of the forest president, shall keep and account to the president for all the instruments and furniture of the school, exercise the closest oversight and watchfulness over the pupils, and he shall be responsible for the practical exercises belonging to the instruction and the exact execution of the work.

Section 27th.—Those who wish to enter as pupils in the forest school ought to hand in to the president about the middle of September an autograph writing applying for admission, accompanied by a certificate from the parish priest to show that the applicant is between twenty and thirty years of age, and is of good repute; also a doctor's certificate as to his good and perfect form of body, together with freedom from incurable disease; also, if the applicant has been previously in yearly service, a proper certificate of dismissal. And the applicant must, at the entrance examination, show himself capable of reading Swedish and Latin printed characters and manuscript; he must write a good hand, and be able to make use of the first four rules of arithmetic. In special cases it may be allowed to the Forest Administration to permit certain exceptions from the rule as to the age of the applicant.

Section 28th.—The pupils must be obedient to the commands of the overseer and president, and must show diligence, order, and a good moral training.

Those who deviate from this must first be warned by the president; and if they do not show any improvement, it belongs to the president to expel them from the school.

ORDINANCES FOR THE PUBLIC SCHOOLS OF FORESTRY IN THE KINGDOM OF SWEDEN.

SECTION 29TH.—The director and the lector in the Forest Institute, together with the president of the Forest School, in case either of them obtains a place as hunting-master, obtains therewith in the class of payments a place corresponding to his year of service, so soon as a vacancy occurs in that class.

Section 30th.—With respect to the appointment of servants by the forest institutions, and to freedom from service for those already appointed, also what relates to mistakes made by servants, is all laid down in the Royal instructions for forest administration and forest law, dated 19th November, 1869.

Section 31st.—With respect to the entrance and leaving of houses by the forest institutions, the same law is in force with respect to the Forest Institute which prevails with respect to like matters in the metropolis. This, together with what respects forest schools, is prescribed in the above-named instructions. With respect to hunting-masters and Crown hunters, that which is there laid down respecting hunting-masters shall be applicable to presidents'; and what is laid down about the dwellings of Crown hunters shall determine the same as to forest overseers.

Section 32nd.—These ordinances shall lie for consideration until the month of October next, at which time the royal ordinances for forest institutions of the 25th May, 1860, cease to be in force; nevertheless, so that all shall begin before the beginning of June next to arrange for the practical preparation which is demanded for the entrance of pupils in the Forest Institute, while in June month of the next year they shall undergo the entrance examination, in the order prescribed in the last-named ordinances.

All whom this concerns have to take note of the same till further advised. We have signed this, and caused it to be confirmed with our royal seal.

C. F. WAERN, *Stockholm Palace*, 15*th Sept.*, 1871.   (Signed) CHARLES.

At the Forest Institute the studies embrace physical science, mathematics, forest engineering, and forest laws. Under the first head is included instruction in mineralogy, geognosy, meteorology, inorganic and organic chemistry, vegetable physiology, botany, and the botany of forests, including plants found in the forests of Sweden, and that of foreign trees, shrubs, and bushes which may be cultivated in Sweden in the open air. In connection with mathematics instruction is given in algebra and geometry, spherical trigonometry, land surveying, and levelling, with exercises in the use of instruments and calculations employed to obtain exact as well as proximate results, and in the preparation of geometrical diagrams, plans, and projections, and the principles followed in giving geometrical descriptions.

The instructions in forest economy embrace—

1. *Sylviculture:* the culture, maintenance, and protection of woods ; and as introductory to this, the leading features of the statistics and history of Swedish forests, and the forest literature of Sweden and of other lands.

2. *Forest Management*, with a special application of this to the forests of Sweden ; and as introductory to this, the leading features and progressive development of the theory and practical management of forests.

3. *Forest Technology :* the utilization of forest products, the felling and transport of wood, the dressing and cutting of timber, and the preparation of secondary products.

4. *Forest Mathematics :* the valuation and calculation of cubic measurements of wood in trees and in woods, natural and artificial, of the age and of the annual increase in cubic contents of trees, and the different methods followed in the valuation of forests, or forest statistics, the valuation of forests for purchase, *excambre* or exchange, division of estates left in heritage, exappropriation, &c. ; forest architecture and engineering, comprising the construction of houses, watercourses or dams, roads, bridges, &c., required by the necessities

of forest technology; and exercises in the solution of mathematical problems to be met with in forest economy, but which cannot be classed under forest management.

5. *Science of Forest Administration :* general basis of the tenure of woods, use to be made of reports of work and results in forest management,&c.; forest book-keeping.

6. *Science of the Chase :* study of Scandinavian quadrupeds and birds, treated as game, and of the methods of hunting them and of the capture of them followed.

7. *Theory of Shooting :* shooting at fixed objects and at objects in motion, and theory of portable firearms.

8. *Cartographic principles* to be followed in the drawing and preparation of forest charts.

9. *Political Economy and Science of Finance,* with a special application of these to forest economy.

Under the head of forest laws instruction is given in financial and economic legislation in their general relations to functionaries, and those employed; in regard to the central administration and local authorities and officials, with whom functionaries and *employés* and forest agents may have occasion to come into communication; in regard to the different categories of woods and forests; in regard to legal partitions of these when required; in regard to the distribution, assessment of taxes, reduction and modification of them; in regard to estimates of contents in cases of exchange or *excambre,* and alteration of boundaries or exappropriation; to duties of drainage and of fencing, and of·taking charge of domestic animals.

Students in the first year of their studies during the autumn session spend weekly in the following classes the number of hours stated :—

Sylviculture	6	hours a week.
Cartography	2	,
Elementary Mathematics	6	,,
Works of Forest Management	6	,,
Structure and Physiology of Plants, and technical uses of trees	8	,,
Chemistry	2	,,
Physical Science	1	,,
Legislation	3	,,
Political Economy	1	,,
In all	35	,,

And in the spring session following, *i. e.*, of the first year of their attendance in the class study of

Sylviculture	6	hours a week.
Cartography	2	,,
Forest Mathematics	4	,,
Principles of Land Surveying and of Levelling	2	,,
Forest Management	6	,,
Forest Botany	4	,,
Ornithology	3	,,
Exercise in Shooting	1	,,
Chemistry	2	,,
Physical Science	1	,,
Legislation	3	,,
Political Economy	1	,,
In all	35	,,

In the second year of their attendance during the autumn session they spend in the class study of—

Forest Technology	6	hours a week.
Cartography	2	,,
Forest Mathematics	4	,,
Drawing of Geometric Diagrams	2	,,
Works of Forest Management	6	,,
Mazology	4	,,
Entomology	4	,,
Legislation	3	,,
Chemistry	2	,,
Physical Science	1	,,
Political Economy	1	,,
In all	35	,,

And in the spring session of this college year in the class study of—

Forest Technology	6	hours a week.
Cartography	2	,,
Management of Forests	4	,,
Forest Architecture	3	,,
Works of Forest Management	5	,,
Entomology	4	,,
Carried forward	24	,,

38

Brought forward . . . 24 hours a week.
Science of the Chase and Exercise in
Shooting . . . . 4 „
Forest Administration . . . 3 „
Chemistry . . . . 2 „
Physical Science . . . 1 „
Political Economy . . . 1 „

In all . . . . 85 „

In the *Journal of Forestry*, vol. i., pp. 756—761, 837—842, are
given the opinions of foresters in Germany in regard to the respective
advantages of having schools of forestry incorporated in Universities
or Polytechnical Institutes, and of having them maintained as separate
exclusively professional schools. In Sweden this subject has not
been ignored.

The National University, deservedly famous both for its past and its
present position among the schools of learning in Europe, is situated
at Upsala, another is at Lund, and there is in Stockholm, near
the Observatory, at the upper end of Drotting Gatan, or rather of
Kungs Backen, which is a continuation of it, a well-equipped
Technological School, and in a separate building by the side of it
a School of Mines—both of late erection; and some saving of
expense and economizing of working power might have been effected
probably by the Forest Institute having been connected with these; but
in all the circumstances of the case it was considered preferable that it
should be continued on its present site and in its existing condition.

In Sweden education is universal. Statistics show that at the
present time even of criminals of all ages only three per cent.—
probably mostly minors—are totally without school training; and the
University of Upsala has done much to secure for all classes of the
community a high degree of education. For admission to study
at the Forest Institute, at the Technological School, at the School of
Mines, or at the University, the requirement is the same. It consists
in the production of a certificate of having passed with approval the
examination in languages, mathematics, natural history, and physical
science required of students leaving the secondary or intermediate
schools.

In the School of Forestry, as in the University and other institutions, no fees are charged; the instruction is free, and bursaries are
provided for those who may require such aid; the board, lodging,
and travelling expenses of the students when they are sent for
practice in the forest are provided for; where steamboats and railways
fail them, ponies are provided if practicable and desirable; their
lodgings and food in these excursions are plain, and in some cases the

latter has to be taken with them or sent on before ; they are exercised in planting, felling, measuring, estimating, and building; and the excursion is enjoyed as a holiday. On completing their curriculum and receiving a satisfactory certificate they are entitled to employment in the Government service, but they are free to accept private engagements. In the public service they have rank corresponding with that of officers in the army, and better pay ; promotion follows according to length of service, but may be anticipated by special appointment given by the Government on recommendation by superiors.

As is the case in the University, the instruction is communicated by lectures and by extempore addresses from notes, definitions being stated slowly and distinctly, that they may be taken down by the students. Repetitions or examinations are held weekly. In these, which are distinct from the daily recapitulation of preceding days' lectures, the students are questioned, but questioned not with a view to ascertain whether they have attended to what was said, but to ascertain whether they have understood this aright,—testing thus the way in which the professor has done his part of the duty, rather than testing the way in which the student has done his part.

In the arboretum the students are made acquainted with the habit of different trees, and they are exercised in nursery work, forming beds, sowing seeds, and transplanting into second beds young seedlings. The students on leaving receive a certificate from the Forest Institute, in which it is certified that he has undergone the required examinations in forest taxation, forest management, forest technology, forest architecture, in the mathematics required in forestry, land surveying, game and forest laws, forest administration and book-keeping, political or national economy, meteorology and soils, chemistry, physics, forest botany and vegetable physiology, forest entomology, mazology, and ornithology, usages in hunting and shooting, and in chart-drawing ; and that throughout the time of attendance at the Institute his application, progress, and conduct have been good, bad, or indifferent, as the case might be. This is signed by the director, the professors of forest economy and mathematics, of botany, and of zoology, and the teachers of game and forest law, of chemistry and physics, and of political economy. There are three degrees of excellence, characterized by appropriate designations, such as commendable, good, and passable, used in the certificate in regard to the different qualifications certified.

About 21,000 kroner (about £1,125) are annually appropriated to meet the expenses of the Forest Institute.

Here, as in Germany, it seems to be acted on as a principle that every one employed in the service of his country should be provided with the best possible education and instruction required for the efficient discharge of his duties; hence the liberal spirit in which instruction is given without any payment of fees.

The Forest Schools spoken of in contradistinction to the Skogs Instituten, or Forest Institute, are dispersed throughout forest districts, of which there are several supported by Government. Besides these arrangements, pupils in the Falk-school, or common schools, receive instruction in horticulture and tree planting. From the report of 1867 it appears that in that year 21,850 pupils received such instruction.

In illustration of what is stated in section 29, it may be mentioned that as in the forest service of India in more than one of the countries in northern Europe, all who go through the prescribed training are entitled to appointments in the service of the State under certain conditions.

Mr. Quin, supervisor of cutters' official measures of timber at Quebec, was commissioned in 1861 to visit Europe to impart and obtain information in regard to the timber trade. He embodied in his report the following statement of the prices, &c., at Swedish ports :—

PRICES OF SWEDISH TIMBER AND DEALS ON THE 20TH MARCH, 1861, FREE ON BOARD.

GOTHENBURG.	Mixed.	Thirds.
Planks, Deals, Battens, and Boards.	£ s. d.	£ s. d.
Redwood, 3 x 11, 3 x 9, 4 x 9, & 2 x 9, per St. Ptg. std.,	8 15 0	7 10 0
„ 3 x 8, 3 x 7, 2½ x 7, & 2½ x 9 „	7 15 0	6 10 0
„ 2½ x 6 and under sizes ... „	7 5 0	6 0 0
„ 1½ x 9 & 8, 1¼ x 9, 8 & 7, & 1 x 9, 8, & 7 „	6 5 0	5 0 0

NORRKOPING, GEFLE, SODERHAMN, LJUSNE, AND PORTS OF SIMILAR PRODUCTION.

Planks, Deals, Battens, and Boards.		
Redwood, 3 x 11. 3 x 9, 4 x 9, & 2 x 9. per St. Ptg. std.,	7 10 0	6 10 0
„ 3 x 8, 3 x 7, 2½ x 7. & 2½ x 9 „	6 10 0	5 10 0
„ 2½ x 6, & under sizes ... „	6 0 0	5 0 0
„ 1½ x 9 & 8, 1¼ x 9, 8 & 7, & 1 x 9, 8 & 7 „	5 5 0	4 0 0

SUNDSVALL, HERNOSAND, NYLAND, AND PORTS OF SIMILAR PRODUCTION.

Planks, Deals, Battens, and Boards.		
Redwood, 3 x 11, 3 x 9. 4 x 9, & 2 x 9, per St. Ptg. std.	7 0 0	6 0 0
„ 3 x 8, 3 x 7, 2½ x 7, & 2½ x 9 „	*6 0 0	5 0 0
„ 2½ x 6, and under sizes ... „	5 10 0	4 10 0
„ 1½ x 9, 8 & 7, 1¼ x 9, 8 & 7, & 1 x 9, 8 & 7 „	5 0 0	4 0 0

SKELLEFTEA, LULEA, PITEA, AND PORTS OF SIMILAR PRODUCTION.

Planks, Deals, Battens, and Boards.		
Redwood, 3 x 11, 3 x 9, 4 x 9, & 2 x 9, per St. Ptg. std.	6 10 0	5 10 0
„ 3 x 8, 3 x 7, 2½ x 7, & 2½ x 9 „	5 10 0	4 10 0
„ 2½ x 6, and under sizes ... „	5 0 0	4 0 0
„ 1¼ x 9, 8 & 7, 1¼ x 9, 8 & 7, & 1 x 9, 8 & 7 „	4 0 0	3 10 0

SUNDSVALL, HUDIKSVALL, NYHAMN, NYLAND, &c.

Timber.	£ s. d.
Best Redwood Square Timber, 9 to 14 in. & up., 30 to 31 ft. av.	1 8 0 per ld.
Best Redwood „ 9 to 13 „ „ 26 & 27 „	1 6 0 „
Red Deals or Battens, for stowage only	6 10 0 pr. std.

SKELLEFTEA, LULEA, PITEA, &c.

Timber.

Best Redwood Timber, 9 to 13 in. sq., av. 20 to 22 feet per pc. 1 1 0 per ld.
Under 9 inches and Whitewood, 5s. per load less.

IV.—FINLAND.

In Finland, as in Sweden, may be studied endeavours to bring the management of extensive forests into accordance with the suggestions of modern forest science crowned with considerable success. But to the student of forest economy Finland also is interesting as supplying an opportunity of studying an antiquated treatment of forests, known in France as *Sartage*, here as *Roehden*, in Sweden as *Svœdanje*, and in India as *Kumari*, &c., which still is practised in some parts of the country, but is being abandoned, and which in India has been much condemned by officials in the forest service.

As St. Petersburg abuts on Finland, and a railway from that city traverses the southern coasts of the Grand Duchy, it may be entered most easily from that city ; or it may be entered by the route which this season I followed, proceeding by steamer from Stockholm to Hango, or to Abo, pronounced *Obo*, or to Helsingfors, and thence by railway to St. Petersburg, the traveller returning, if so disposed, by steamer to those places and Stockholm.

I arrived too late by an hour to take the train from Abo, and again too late by two hours to take the day train from Helsingfors. I kept by the steamer, and proceeded by her to St. Petersburg in preference to travelling by a night train.

In travelling by railway the traveller may at different stations penetrate some way into the country by branch lines. Some of the interesting localities which may be thus visited I had visited before ; with others I had been familiar for many years.

In the *Journal of Forestry*, vol. i., pp. 545—551, 701—705, is an account of the Finnish School of Forestry, which is situated at Evois.

From Rilhimski a branch line goes to Tavastehuis, which may be reached in three hours from Helsingfors, or in longer time from either of the other ports named. Thence a journey of some fifty versts (or thirty-three miles) by chaise—the one-horse conveyance of the country—will take the traveller to Evors. The journey will occupy about six hours. It leads through the post stations of Heinökangar, Syrjäntaka, Yso-Eve, and Evon Opista. The charge is ten penni per verst, or one mark, equal to tenpence sterling, per Finnish mile of ten versts, or between six and seven miles English. At the hotel (Hôtel Nordin) every assistance will be courteously given to the traveller.

At Wyborg there may be found facilities for visiting the famous Falls of Imatra. These I visited many years ago, long before these facilities were created; and I have given an account of my visit to them in a volume on the "Hydrology of South Africa," pp. 24—26.

In the issue of the *Queen*, the Lady's Newspaper, for October 12 last, is a cut representing a sketch of the falls, but one which I would never have recognised as such. It is accompanied by the following account of them :—

"Firland, with its boundary of bold and precipitous granite cliffs, and its magnificent forest, lake, and mountain scenery, may be called the Switzerland of Scandinavia, and is a favourite resort for Russian tourists during the short but genial northern summer. One of the most frequented sights are the Imatra Rapids, which have been made easily accessible by a company which provides every accommodation for travellers. The rapids are formed by the river Wuoksi, which joins the Lake Saima with Lake Ladoga. On its way it has to traverse the mountain side of Salpansfelka, in a furrow 140 feet wide. Through this narrow channel of about three thousand feet length, with a fall of sixty-three feet, sixty-seven millions of cubic feet of water force their way every hour; and the grandness of the sight may be imagined when we learn that within the same time the Niagara Fall sends only about forty-two million cubic feet into the basin below.

"High above the present bed of the rapids the roaring waters have hollowed out immense caverns in the adjoining rock, which show the progress of the river tunnelling from immemorial times. The communication between the two banks is effected by means of a wire rope, 170 feet long, and only three and three-eighths of an inch in diameter, along which a basket runs, with room for two passengers. On the spot where the rush of the waters has attained its greatest velocity, a pavilion erected on a rock affords shelter to enjoy the sight in comparative comfort."

A journey of four hours by rail will bring the traveller from Wyborg to St. Petersburg.

A perusal of what has been stated in the article on the School of Forestry at Evois, which appeared in this Journal last year (vol. i., pp. 545—551 and 701—707), may suffice to give some idea of the present state of forestry in Finland ; and therefore here there is brought under consideration only the practice of *Roehden*, as it is there called, that of burning a portion of the forest for the *temporary* culture of cereals, differing thus from the extensive clearing away of forests for permanent occupation and agriculture. It consists in simply cutting down the trees on a spot which the people design to cultivate, and burning upon this the felled trees, that the ground may be fertilized with the ashes. The ground is then sown with rye, barley,

&c., or, if very fertile, with oats and buckwheat, after which it is abandoned, to be again covered with forests, or to lie waste. The evils produced by this course of procedure will afterwards appear. In another connection I have stated:—

From what is said by W. von Schubert in his " Resa genom Sverige Norrige, Lappland, &c.," published in Stockholm in 1823, in 3 vols. 8vo.; and from what is said by Lars Levi Laestadius, in his work entitled " Om Möjligheten och Fördelen af allmänna Uppodlingar i Lappmarken," published in Stockholm in 1824, it appears that the practice of burning over woodland at once to clear and to manure the ground, and from other incidental references to it, is still a recognised usage in Swedish husbandry.

There it is known as *Svœdanje*, which Swedish designation is also in use in Finland, which was formerly a Swedish province. Though used in Norrland in Sweden as a preparation for crops of forage or grain, it is employed in Lapland more frequently to secure an abundant growth of pasturage, which follows in two or three years after the fire ; and it is sometimes resorted to as a means of driving the Laplanders and their reindeer from the vicinity of the grass grounds and the haystacks of the Swedish backwoodsman, to which they are dangerous neighbours. The forest rapidly recovers itself, but it is generally a generation or more before the reindeer moss grows again. When the forest consists of pine (*tall*) the ground, instead of being rendered fertile by the process, becomes hopelessly barren, and for a long time afterwards produces nothing but weeds and briars. I have elsewhere had occasion to remark,—

" It is a practice," says M. Parade, formerly Director of the School of Forestry at Nancy, " *extremement ancienne*." And such it appears to have been in France; but there may be claimed for it an antiquity far greater than is indicated by the practice of it in France, in Sweden, or in Finland; and amongst the conservative tribes of India it has been practised to an extent which makes the *sartage* of France, the *roehden* of Finland, and the *svœdanje* of Sweden appear as mere childish play. In the Canara district it is known as *kumari*. In a document issued by the Board of Revenue in India, in 1859, it is stated that, in some parts of Bekal, which is the most southerly of the taluks of Canara, kumari cutting forms part of the business of the ordinary ryots, and as many as 25,746, or one-sixth of the population, are supposed to be engaged in it; but to the north of that taluk it is carried on by the jungle tribes of Malai Kadeos and Mahratas to the number of 59,500. Here we have upwards of 85,000 men felling, burning, and destroying forests, for the sake generally of one—or at most of two crops—sometimes, but rarely, of three. After which the spot is deserted until the jungle is sufficiently high to tempt the kumari cutter to renew the process.

By this practice vast quantities of most valuable timber have been destroyed.

A good crop of hill rice, or Nullet, is obtained in the first year after the consumption of the wood, a small crop is taken off the ground in the second year, and sometimes in the third, after which, as has been stated, the

spot is deserted. In the south, where land is more scarce compared
with the population, the same land is cultivated with kumari anew in
12, 10, or 7 years; but in North Canara, the virgin forest, or old kumari,
which has not been cultivated within the memory of man, is generally selected
for the operation.

'This rude system of culture,' says Dr. Cleghorn, formerly Conservator of
Forests in the Presidency of Madras, 'prevails under various names in different eastern countries. It is called *kumari* in Mysore and Canara, *pounam*
in Malabar, *punaka* in Salem, *chena* in Ceylon.'

The name *kumari* is peculiar to the Canara and the Mysore districts. It
is thus described in an extract from the Proceedings of the Board of Revenue:
—'The name is given to cultivation which takes place in first clearings. A
hill-side is always selected, on the slopes of which a space is cleared at the
end of the year. The wood is left to dry till the following March or April,
and then burned. In most localities the seed is sown in the ashes on the fall
of the first rains, without the soil being touched by implement of any kind;
but in the taluk of Bekal the land is ploughed. The only further operations
are weeding and fencing. The crop is gathered towards the end of the year,
and the produce is stated to be at least double that which could be obtained
under the ordinary modes of cultivation.'

Gabriel Rein, in a volume entitled *Statistick Techning äf Stor-
furstendömet Finland*, third edition, 1853, states that from very ancient
times the Finns had practised agriculture, and for centuries this
has been their chief means of support; and this more so as the
supplies derived from hunting and fishing became diminished.
It is to the honour of the Finnish people that by them agriculture
has been carried to the most northern boundaries of their country,
and has thus elevated or prepared for a higher state of civilization the
inhabitants of those regions. And there it may be seen practised by
the most northern agriculturists about Altengard in Norway, where
the Finns are called *Quains*, the designation Finn being there given to
Laplanders.

The soil of Finland is naturally unsuitable for a highly developed
system of agriculture, partly in consequence of the stony character of
the ground, partly in consequence of the poor character of that
soil requiring the labour of years to fit it for superior culture, and
partly in consequence of the sparseness of the population; and such
was everywhere originally the condition of the country. So far back
as history reaches, the Finnish people have carried out this kind of
nomadic use of the soil, mode of culture which could be applied
for a short time to one portion of the land, and then, this being
abandoned, be applied to another.

The designation *Svedanje* is derived from *Sveda*, to scorch or burn,
or burn off. This term and the corresponding term *Svedja* seemed to

be applied exclusively to this mode of forest clearing with a view to temporary agriculture. In the recent cases of this all the trees growing on a piece of ground are cut down, allowed to dry, and burned. The field is then ploughed, or rather scratched with a rude harrow, whereby it is loosened and intermixed with the ashes. It is next sown with rye, barley, or other seed. If the earth proves somewhat fertile, it is sown next year with oats, and it may be afterwards with buckwheat. After this last crop has been reaped, the ground is abandoned and left to rest till it be again overgrown with trees.

Ofttimes it is the case that through the culture and removal of these crops the soil becomes so exhausted that for many years no trees are produced, and being used for pasturage, the cattle not only grazing on the selfsame grass, but browsing on seedling trees as they may make their appearance, the growth of these is entirely prevented. The forests have thus been greatly diminished, and the products of the forest being of great importance to Finland, the Government has endeavoured to limit as much as possible this mode of culture. In consequence of this, and the introduction of improved agriculture, *Svedja* has been greatly diminished. It is practised now to a very limited extent in the län or district of Nyland, in which Helsingfors is situated, and that of Abo, and throughout the greater part of Tavastehuis län to the north-east and north of these districts, and of Wasa, on the coast of the Gulf of Bothnia, and the greater portion of the Uleaborg län. It may be said to be in these districts practised only occasionally, as on the uprooting of clumps of trees growing in the middle of a field or meadow, or with a view to preparing pasture land for cattle. Consequently the product of *Svedja* in these districts is small, very small in comparison with that obtained by the usual mode of agriculture. *Svedja* is not practised now in the eastern läns, except in those läns near the coast and in the south, where, for example, in the south-west of Wyborg län it is only occasionally practised; and at Borga, Wyborg, and Nykyrka it has been altogether abandoned. On the shores of Lake Ladoga it, is strictly prohibited, in consequence of the scarcity, of wood. In most other parts of the län where it is still practised this is done as a necessary and at present unavoidable means of support. The produce of it in Säckjärvi is in proportion to that otherwise obtained as 1 to 3. In other places it is as 1 to 2. In St. Michael's further to the west it begins to be limited. The poverty of the land and the limited produce yielded by regular agriculture lead to the greater use of *Svedja*.

It is in consequence of the injurious effects of the practice that it has been thus limited.

This mode of culture has been observed to impoverish the soil so much that the forest destroyed is not always replaced by a new growth, and now only in districts where the uneven and stony condition of the ground renders it difficult to carry out regular husbandry does this mode of culture still prevail.

Some of the evils resulting from the similar practice in India, known as *Kumari*, are detailed in reports embodied in the volume entitled, " The Forests and Gardens of South India," by Dr. Cleghorn. There are two forms of it which have been utilized in France, known respectively as *Sartage à feu courant* and *Sartage à feu couvert*.

" Apart from the destruction of the trees and the laying bare of the soil," says Marsh, "and consequently the free admission of sun, rain, and air, to the ground, the fire of itself exerts an important influence on its texture and condition. It cracks and even sometimes pulverizes the rocks and stones on or near the surface; it consumes a portion of the half-decayed vegetable mould which seemed to hold its mineral particles together, and to retain the water of precipitation, and thus it loosens, pulverizes, and dries the earth; it destroys reptiles, insects, and worms, with their eggs, and the seeds of trees and of smaller plants; it supplies in the ashes which it deposits on the surface important elements for the growth of a new forest clothing, as well as of the usual objects of agricultural industry; and by the changes thus produced it fits the ground for the reception of a vegetation different in character from that which had spontaneously covered it. These new conditions help to explain the natural succession of forest crops, so generally observed in all woods cleared by fire and then abandoned."

V.—NORTHERN RUSSIA.

THE routes to St. Petersburg from Britain are numerous. During the summer season there is scarcely a day in the week excepting Sunday on which a steamer is not leaving some British port for St. Petersburg in times of settled peace, and the voyage is always made within the week; and by railway and steamer it may be reached in less time by Ostend, Antwerp, or Rotterdam and Cologne, by Hamburg and Berlin, or by the route through Sweden and Finland.

A simple division of Russia into well-defined regions has been adopted or devised by Mackenzie Wallace, and in accordance with this he has prefixed to the second volume of his work on Russia a map of Russia in Europe, in which he represents what he calls the Forest Zone as extending in width from the whole breadth of Finland, and following to some extent the 60° parallel of latitude, the latitude of St. Petersburg, to the Ural Mountains, and bounded on the north by Lapland, the North Sea, and the country of the Samoyedes. To the south of it is the northern agricultural zone, extending from the Baltic, Prussia, and Poland, to and a little beyond the Ural Mountains, contracting considerably in breadth and stretching more northwards in its eastern half. To the south of this is the southern agricultural zone, or black earth zone, beyond which is the steppe zone; and beyond this to the south-east is the pastoral region, to the north of the Caspian, the whole spreading in a fan-like figure as from a pivot in the latitude of 55° N. in Asiatic Russia.

In these different regions of the extensive Empire of Russia, compared with which the whole of Western Europe may seem like a province, may be studied in operation all the varied methods of exploitation from their first development to their most advanced form. *Roeden*, as it is called in Finland, the *Kumari* of India, the *Sartage* of France, is also to be met with in the north, but this can scarcely with propriety be designated forest exploitation. But there we find dominant the primitive form of forest exploitation,—that known in France sometimes as *Furetage*, but more generally as *Jardinage*, a mode of exploitation followed in several British colonies, and entailing upon them, I fear, most disastrous consequences.

The designation *Furctage* has been derived from *furet*, a ferret, and it has been given, I presume, in allusion to the forester seeking out or ferreting out the tree he requires, and felling it alone; but it is more generally called *Jardinage*, a word derived from *jardin*, a garden. This designation may have been given from the similitude of the operation to that of the gardener who gathers here and there the pot-herbs most advanced in condition, or in immediate danger of going to seed; and when carried out extensively and long it produces in the forest an appearance not unlike that of the half-cleared kitchen-garden, in which may be seen bean-stalks, void places, cabbage stocks, half-reaped beds of turnips, of carrots, of onions, with withering pea straw hanging from stakes inserted for its support when it was fresh and young, and weeds of various kinds growing between. This mode of operation is not confined to extensive forests, though practised in these extensively in Russia and in America, as well as in Britain and British dependencies. It is one well known to all foresters who have to manage ornamental plantations, in the removal from these of trees showing symptoms of decay, or of others in a good growing condition deemed suitable for the market or required for use and doomed to the axe, while the trees growing around are spared.

In the central zones we meet with a more advanced form of forest management, and in the southern zone we meet with forest extension carried on to such an extent as to give a character to the forest economy of the region.

It is to the forests and forest economy of the northern region alone to which attention has now to be directed.

I should have enjoyed having an opportunity personally to visit this region with the operations on which I have long been acquainted; but while I was inquiring about the arrangement of steamers for Lake Onega, Professor Schafranoff put into my hands a printed narrative by Mr. Judrae of a journey which he made to and beyond that region, with qualifications and facilities for obtaining information far greater than mine, which rendered my journey unnecessary. Of part of this narrative the following is a translation :—

"The first steamer of the season (1867) proceeding from St. Petersburg to Petrozavodsk, sailed on the 30th May (Old Style), having been prevented from sailing earlier by the ice on the Neva and Lake Ladoga. With fine, somewhat warm weather, we left the capital, and a few hours' hard steaming against the current brought us to Lake Ladoga; but scarcely had we got 30 versts (twenty miles) from St. Petersburg when ice began to meet us, some of it in sheets of a very large size; and it was getting dark. The keen north-east wind made itself felt; and looking to the horizon there stretched out before us a sea of unbroken or of congealed fields of ice; the steamer, however, resolutely advanced. I took refuge in the cabin from the intolerable cold, but after a few minutes I hastened on deck in consequence of the steamer being 'stopped. There was ice in immense shoals ahead of us, so that to go on in the course we were following would have risked damage to our paddle-wheels, whereby we should have been placed in an awkward condition amongst the ice floes of the Ladoga. At length the order was given to cast the anchor and wait for the day. In a few minutes we were fast, and a strangely contrasting stillness and silence pervaded the vessel, while a magnificent scene was stretching around us in all directions. Far as the eye could see were open spaces of water and sheets of ice commingled, and whole schools of black seals moving backward and forward on the floating masses, while with the cold wind were combined black clouds and a murky sky, although it was now the 31st of May (O. S.), the 12th day of June in lands where the New Style has been introduced.

" Next day, the steamer by some way or another got through Lake Ladoga, and entered the river Svir._ Steaming along we found everywhere on the banks on both sides, woods, woods, woods. From the deck of the vessels could only be noticed firs and pines and birches, although in some parts of the government of Olonetz there still grew the Norway maple, the lime, the elm, and other kinds of trees.

" Now we passed on the left bank of the river the town of Lodenoi-Pole, founded by Peter the Great, and formerly a naval dockyard. A few hours more and we reached Vosnecenya, one of the principal centres of inland navigation by a system of canals, of which there are two or three connecting the Volga with the Baltic.

"The village of Vosnecenya is situated on the Svir as it issues from the Lake Onega, and it is called by the inhabitants the Petersburg Gate.

"It was impracticable to go further by the steamer, as the ice in this lake had not yet broken up; consequently I had to travel to Petrozavodsk by horse, which I did by a very picturesque route by the western shore. From Vosnecenya to Petrozavodsk by the so-called Vilegarskoi road is 130 versts, or 86 miles.

" The first thing which interested me was the Forest-Product Manufactory of Mr. Baelaeff, well known in all these northern parts. It is situated about seven versts from Vosnecenya.

" The lovely view presented by the Fabrique and the buildings around leads me to conclude that it must be a profitable property, yielding a considerable revenue. It is built in a situation very convenient for the sale of the products; near to Vosnecenya, where there is a great consumption of tar in caulking vessels. Hitherto there could only be obtained black burnt tar, which is not quite suitable for the purpose, and the demand for it was not great; but now they are constructing new brick furnaces for the production of what is called red tar, from the sale of which they will, without doubt, obtain considerable profits.

"There is not a scientific or special manufacturer employed, but the works are under the management of an able workman ; by this arrangement it is supposed a great saving is effected. The Fabrique contains at present several furnaces, by which are obtained tar, turpentine, and other products from pine wood. Besides these, there are furnaces for rectifying turpentine and for making pyroligneous acid. The latter product is obtained from birch wood by a process of dry distillation. The greater part of this product is taken to St. Petersburg, but the greater part of the red tar commands a sale in the locality.

"The following are details relating to the manufacture of such articles obtained on the spot. From a cubic fathom of wood are obtained 25 poods of black tar, equal to 900 lbs. English, and two poods or 72 lbs. of turpentine ; and in the manufacture they consume half a cubic foot of firewood.

"From a square fathom of birch they obtain 250 [?] poods of pyroligneous acid.

"How far these figures indicate the reasonableness and propriety of the measures adopted may be determined by a comparison of them with results obtained by scientific operations, and with the returns made by other works of the same kind elsewhere. The proprietor was desirous of impressing on me that the establishment is not remunerative, and hardly returns the working expenditure. The quantity of acid manufactured is some hundred tons more than suffices to meet the demand for it, and the turpentine is scarcely equal in quality to what is required in the market, and thus he accounted for the unremunerative character of the works.

"Between the hills are occasionally met with rivers or rivulets flowing into Lake Onega. The current of these is very rapid in consequence of the steep declivity of the ground towards the lake ; and they present generally the characteristics of mountain streams. The most striking feature of the country is the great quantity of boulders upon its surface, the number of which, if stated, would be almost incredible. They consist exclusively of granite and other primary or transition formations, covered partially by drift, in which is a red sand in considerable quantity. There are also projecting from the ground granite hills in whole or in part quite bare, or covered only with lichens. Having examined the works I proceeded further. On the right hand was a magnificent view of the Onega lake, the breadth of which at this place is above 80 versts (about 54 miles). On the left side of the road were hills, the continuation of those of Finland, which pass into the government of Olonetz, but fall away towards the south till they present an altitude not exceeding 420 feet above the level of the country around. The surface of the hills seen from this point is covered with forests, which consist of four different kinds of trees, intermixed in varying proportions—fir, pine, birch, and aspen. The height of these trees, judging by the eye, seemed to be low compared with like vegetable productions. The only impression I have retained of the course of the whole journey to Petrozavodsk with such opportunities of observation as I had, was a feeling that I had gotten into a comparatively northern region, and that I must be nearing the polar circle ; granite hills and interminable forests, a stony soil, with abundance of waters but a sparse population—these are my remembrances of my first acquaintance with the government of Olonetz.

" Every twenty or thirty versts (14 and 20 miles) there were small villages inhabited by Karrells, a tribe of Finns who have retained the Finnish language, but in every other respect they are like the population from Novgorod found in the south and east and in parts of the central portion of this government.

" Speaking generally, the first acquaintance of one with the country leads to the conclusion that the government of Olonetz is as poor in works employing human industrial labour as it is rich in natural productions, amongst which the first place must be assigned to those of the forests. In Petrozavodsk I was enabled to collect from records by officials who had formerly the management of the forests, and of all matters relating to the country, information of which the following is a summary.

" The government of Olonetz lies between 60° 21' and 65° 16' N. Lat., and 47° 21' and 59° 36' E. from the meridian of Faro, corresponding to about 30° and 40° E. of Greenwich. It has an area of 2,785 square geographical miles, or 14,026,320 *desatins.* Of this area forests cover approximately ten millions of *desatins,* or five-sevenths of the whole. After deducting 257,000 *desatins* of arable land, and 88 *desatins* of pasture land, the rest is composed of rivers, lakes, swamps, and other unproductive places. The whole population, including both sexes, amounts to 301,290 ; consequently there is for each man 47 *desatins* of surface, consisting of—

1·14 *desatins* of arable and pasture land,
35·19 of forests; and
10·67 of lake and river.

" This proportion of the population to the area is indicative of the poverty of the territory. The forests belonging to the Imperial Domaines measure 8,774,419 *desatins* and 1,048 square fathoms, or about 1,740 square geographical miles, which amounts to about two-thirds of the whole area of the government. In 1865 the revenue derived directly from the forests amounted to 327,993 roubles,* and by extra fellings 9,607 roubles 90 kopecs; in all 337,540·90 roubles. According to calculation each *desatin* on an average yielded a revenue of 3·84 roubles. In subsequent years the revenue was considerably diminished in consequence of the saw-mills not working.

"In so far as forests are concerned, the importance of the government of Olonetz is seen more in view of the future than in relation to the present. Having several navigable outlets, it may be considered a reserve of forests available not for Russia only, but for Europe.

"Looking into the accounts of revenue derived from these forests, we find that almost 45 per cent. of the revenue is the proceeds from the sale of timber taken to the saw-mills. In 1865 there were sold to seven of these 237,783 logs for the sum of 98,359 roubles 59½ kopecs, and to the English Onega Company, having its *fabrique* on the River Onega in the government of Archangel—but preparing at the present time forest material in the district of Kargopole in the government of Olonetz, logs amounting in value to 52,585 roubles; in all 150,944 roubles 59½ kopecs.

"From what has been said it follows that the saw-mills, which are the principal purchasers, are indispensable for the sale of timber; and that but for these there would be but a small sale of timber, more particularly in the northern parts of the government.

"Reading the reports in the Government office of the Imperial Domaines, one is arrested involuntarily at a place which treats of unauthorized fellings carried on without leave or sanction.

"According to these reports the population of the government consists almost exclusively of those who were Crown serfs and their children, whose requirements of wood for fuel and building are sufficiently met by the allotments made to them annually from the forests; but these people for a long time back have been possessed with the idea that woods are of no pecuniary value, and they destroy them recklessly. When the annual allotment happens to be less than they think they require for building material—for it may be fancy erections which they do not require—they frequently go off to the woods and cut what they want without ever applying for permission to do so. And then the question comes up, Is it possible for the people to acquire at the present time any adequate idea of the

* The standard equivalent of the rouble is 3s. 4d. It is generally, in accordance with the rate of exchange, 2s. 6d. When I was in Russia last year it was 2s., and at one time during the war it was 1s. 10d. The rouble is worth 100 kopecs.—J. C. B.

necessity which there is for the conservation of the forests and the exploitation of them in a rational or scientific way? Let any one realize the case. Around all of these villages, even the smallest of them, there are forests of which the eye can see no end, they appear to be interminable; and there are depths of them to which the foot of man has never penetrated. The extent of these forests is such that to the peasantry they seem inexhaustible; while, on the other hand, the severity of the climate, the unproductiveness of the soil, and the poverty of the people are such as to seem to call upon every one to find out for himself with a hatchet in his hand any means of improving his condition.

"The natural condition of the country could not have called forth or exercised upon the people an effect more to be deplored.

"The peasantry here look upon wood as being in common with earth and air, fire and water, one of the elements, and as equally free to all persons; and they consequently consider that they are free to use it without stint or limit, as one of the free gifts of nature. This state of things, originating, as I have intimated, from the physical condition of the country, can only be changed or destroyed by the great change-producer, time; and the reports of the consequent destruction of the forests embrace numerous details of the extension in the country of the practice of *Sartage* and *Roeden*, or *Svedja*. This system of felling is very frequently met with; but if we enter into the circumstances of the case, considering, on the one hand, the condition of the agricultural economy of the people, together with the paucity of labourers and the lack of manures, and the circumstance that the temporary culture of the fields which is thus effected supplies the only means of support to man, and, on the other hand, the great extent of the forests and the difficulty of maintaining an efficient watch over them by wardens or forest watchmen with a great extent of forest entrusted to their care, we cannot condemn the Forest Administration for not adopting effectual measures to prevent altogether this unauthorized felling of trees in the forest.

" This unauthorized felling is the primary form taken by agriculture —the first step taken towards the development of rural economy. We hope in process of time to get beyond this; but to put it down by force would not be a rational course of procedure. The Northern peasant not having productive ground near his residence, nor means to improve it if he had, goes into the depth of the forest, burns down trees, and cultures his temporary field for two or three years, or so long as its power of fertile production is not exhausted—the fertility being produced by the ashes and cinders of the burnt trees. The persuasion of the peasant as to the perfect legality of such a procedure is such, that it is very doubtful whether any general measure of repression at present could remedy the evil. In order fully to understand the economic condition of this region we must go back some fifty years or so, and look at things with other eyes. I consider that this unauthorized felling originally was legal and reasonable—suitable for the place where the forests are very dense ; but as a principle it admits of some formal limitation. And this, according to these reports, appears to have been attempted in the government of Olonetz in 1867. Of the system of operations carried on by this people, it is said the first settlers in the country were satisfied with small plots of ground of easy cultivation, but as they increased in number they were obliged to have recourse to land which was more fertile indeed, but marshy or covered with forests, and requiring labour to prepare it for culture, and care and thought. Cultivation such as may be seen in civilized communities was not attainable by these people, were it only from their want of agricultural implements and manure. In the same book, on the page following, it is stated, 'In these virgin soils, previously covered with forest or bush, the produce of rye in the first year was *ten*fold—frequently *twelve*fold ; and there were places— generally places where there had been old dense high forests—in which the produce was *fifty*fold, and in the second year the produce was from ten to fifteen fold.'

" Within two weeks after my arrival at Petrozavodsk I was once more on the road in my *kibitka* speeding onward to the most northern town in the government of Olonetz, where according to the opinion among the population is the end of the world. This town is called Povonetz. At about 20 versts, or 14 miles, from Petrozavodsk is the village of Thouya, the first post station on the river of the same name, across which there is a barge ferry. The river Thouya flows into the Onega Lake, and has throughout its course a very rapid current. Where I crossed there was wood being floated down from the Government mining forest estates situated further up, from whence the strength of the current brought them down.

" The current brought them with such rapidity and force that the barge was in danger, and with difficulty we reached the opposite shore.

" The rapid current is not favourable for the flotage of timber, and there has been formed what may be called a dam at the mouth of the river ; but this having been broken, a great quantity of wood has been carried out into the Onega Lake, whereby the navigation of it in this part by steamers has been impeded. It is to be desired that some effective measures were taken to prevent this loss, which increases the cost of what forest timber is secured.

" Looking at the floating timber, I was struck with the activity with which the men employed maintained their footing, each standing on a log and holding in his hand a long pole or boathook, with which he balanced himself, and with which, in floating down the timber, he cleared the obstacles encountered; and these on this river are very numerous.

" For this purpose it is generally inhabitants of the district who are employed, these being very skilful and accustomed to the work. They are here known as ' Onejan,' or Onega men, and I am under an impression that under this general name such workmen pass in St. Petersburg.

Proceeding onward to the north, on both sides of the road there were to be seen forests and forests, and nothing but forests. I can affirm that the person who is acquainted with the extent of these forests by knowing only the number of *desatins* which they cover, has no idea of what that extent is. To obtain this, one must travel through them—travelling continuously through forests for five hundred versts ; and he must experience personally the depressing influence produced by the forests and forest-covered mountains of this forest region to enable him even partially to comprehend what is implied in the easily pronounced statement about so many millions of *desatins*. Such numerical statements are required for the production of a *national tax*, or estimate and prescription of what fellings should be made to secure a sustained production of wood, and charge to be made for trees ; and the latter is a matter which is not so easy of accomplishment as to many at first sight it may appear to be. Those who are in the trade do not make known what is the cost of preparing the timber for the market, or the prices obtained by them, being afraid of the charge to them being raised. If there be made but a simple allusion to the subject, they begin to complain that they are carrying on their operations at a loss, and that the demand for timber is diminishing from year to year. And to arrive at a knowledge of the truth, the forest officials must solve the problem for themselves, with such data as they have at command.

" In connection with this subject the following statement may show approximately what are the proceeds of the sawing of timber. From four logs are produced three dozen of boards of different measurements. Four logs, according to the present charge, cost 1·80 rs.; the transport to the river and flotage, sawing, shipment, and freight to Cronstadt of these cost 10 rs. ; so the total cost is 11·80 rs., and the three dozen boards at Cronstadt are worth 18 rs. But the calculation, it must be borne in mind, is only an estimate approximately correct. The total number of saw-mills in the government belonging to private individuals is *seven*. Of these, two are in the district of Olonetz, two at Petrozavodsk on the rivers Souna and Leejma, and three in Povonetz on the rivers Oumentza, Koumsa, and Povetchanka. The first-mentioned two are at the present time doing nothing, and this is the cause of the considerable diminution seen of late years in the revenue. Of the operations at Petrozavodsk some account has been given.

" At a distance of ninety versts, or sixty miles, from Petrozavodsk, is the village of Leejma, where there is a saw-mill of considerable magnitude, occupied also at the present time by M. Baelaeff. It is erected on the river Leejma, and has two water-wheels and four frames of saws, two for each water-wheel. It works without intermission day and night, and can cut up in the course of the year 60,000 logs ; but, in consequence of hindering circumstances, it cuts up only some 45,000. These are pine wood logs of the length of twenty-two feet, and eight verschoks or fourteen inches thick at the upper extremity. The boards most in demand in the market are twenty-two feet long and three inches thick, which are known as $2\frac{1}{2}$-in. boards ; and besides these there are what are called inch boards, sent chiefly to Holland. According to the statements of the traders these inch boards are both in quality and price inferior to the Swedish boards of the same measurement, in consequence of which the preparation of them in large quantities is not remunerative.

" Coming next to those connected with Povonetz, I have to state that not far from the post road on the river Koumsa, at a distance of twenty-three versts from Povonetz, there is a saw-mill belonging to the timber merchant Mr. Zachanieff. This mill also I had an opportunity of seeing. It is built in a very pretty situation, in the valley of the rapid river Kamsa, surrounded by lofty hills extending to the Onega Lake. The mill has one wheel and two frames, and there are sawn in the course of the year about 30,000 logs. Everywhere about it are seen order and cleanliness ; and there is a fire which never dies out, burning continuously the outside slabs, the ends of logs, and other *débris ;* and what are literally mountains of sawdust fill up the picture of the mill and its surroundings, while the noise of the wheel and of the saws is reverberated by the surrounding forest.

" A journey of some fifteen miles brings us to Povonetz. A poorer and more unattractive town than this it is impossible to imagine : it is simply a village built on the plan of a town. The most remarkable object in Povonetz is an old wooden church standing on the shore of Lake Onega, built by Peter the Great, the only monument which indicates that ever he was here. There is, it is true, besides this, the Petrozavodsk road ; but this is now only a footpath or track, by which are brought the goods obtained in this town from Archangel. Add to this two or three legends or traditions about Peter, and all records of his having been here are exhausted.

" Almost close to the town, on the estuary of the Povetchanka, is the saw-mill, which gives some little life to the town, and is the only thing which vivifies its existence.

" The whole biographies of the place tell only of what relate to the works, besides which the inhabitants have an opportunity several times in the course of the summer to admire a steamboat which visits the place ; but beyond this and fishing, change they have none.

" Almost all the vessels which leave the landing-place of Povonetz are laden with boards produced at this mill. In the fullest sense of the word, Povonetz is a timber town, and on arriving here I felt proud while I thought that my profession was the principal profession of its inhabitants, and had to do with the very source of its wealth. To determine and specify what is the trade of the place must occasion no difficulty to any one. Its imports consist of everything excepting wood and fish, and its exports consist of wood and fish alone, the latter principally Triska.

" The discharge of my official duties led me further in the north.

" For nine versts or six miles beyond Povonetz it is possible to travel by coach, but beyond this point the journey had to be made by water in very uncomfortable boats on narrow lakes, and rivers connected with them. From the Lake Volozer issuesthe river Povet-chanka, which flows through a very picturesque country. Thanks to the high hilly shores, the general rapid current of the river, and the frequent occurrence of considerable rapids, this little river, or rivulet, is in spring changed into a very dangerous torrent, tearing along, and threatening to engulf and carry along with it whatever may tumble into its waters. It has a course of about eleven versts, nearly eight miles, and by it are floated some 20,000 logs a year to the saw-mill at Polonetz.

" The construction of a road from near the Lake Volozer to the White Sea has been projected, and the initiative of the execution has been taken, but nothing more seems to have been done. The proposal created great excitement throughout the district, where there are very few roads of any kind or other facilities for communication with other parts. Scarcely could the projection of a railroad in any other part of Russia produce so much discussion and excite so many hopes as would the making of a common road in this country. This part of the government of Olonetz is passing through that period of its history at which any measures taken for the formation of roads, the opening up or clearing of forests, or the introduction of regular systematic agriculture, possess very great interest.

"Unhappily the execution of this enterprise has not proceeded further than the felling a strip of trees through the forest along which it was proposed that the road should be made. And the general impression is that soon the whole matter will end, for money is not forthcoming, and the kind of road is not satisfactory. Coming upon it at various points, it seemed to me that the projector or surveyor had of design made it to pass at a distance from the most important centres, and carried it over uninhabited districts and unsuitable land.

"For forest operations this road to the White Sea would not have been unimportant, and, having referred to the subject, I am led to mention also a proposal which has been made to open up water communication between the White Sea and the Onega Lake. Having no accurate data, but only partial information, I cannot give details or discuss fully the importance of this gigantic project.

"Of this proposal it is stated in the *Pramiatnais Knjka*, or official Notes of the Government of Olonetz for the year 1867, ' The execution of this project, opening up communication between the White Sea and the Gulf of Finland, and *vice versa*, proposed solely with a view to commercial enterprise, would for strategical purposes affecting the whole of the north of Russia have immense importance ;' and Mr. Seederoff [a gentleman well known throughout this region, a merchant who has carried on great commercial transactions in Archangel and in Nova Zembla, and made valued contributions to the different International Exhibitions in the capitals of Europe] says in a communication to the Imperial Free Economical Society, ' Steam war vessels could proceed from Cronstadt and make their appearance for the protection of the inhabitants of the shores of the White Sea, or, if necessary, of Archangel, which now, in consequence of the dismantling of the fortress of Nova Dwina, is left without defence.'

"According to the views of M. Seederoff there will only be required the construction of a canal fifty versts long, which, opening on the lake, will make it possible for shipping to pass from the lake to the White Sea, or from the White Sea to Lake Onega, and, consequently, to St. Petersburg."

M. Judrae goes on to say, "Mr. Seederoff has, I think, neglected to take into account the rapids of the Svir, which, to the accomplishment of such a scheme, would require to be passed by a canal ; and this would add considerably to the difficulty of the undertaking. But both the Onega and White Sea Canal and the White Sea road remain at present within the category of projects, and they are likely to remain there for some time, as no one seriously believes in the execution of either of them in the immediate future.

"Returning to details of my journey. After proceeding some eighteen versts, or twelve miles, by boat through a succession of narrow lakes, I landed at a place where there was a very narrow path, which could only be traversed on foot. A walk of six versts, or four miles, brought me to the village of Morskoy Mosselgie. The road I found pleasant. It goes along a picturesque ridge of hills, running from west to east some thirty-two versts or twenty-one miles north of Povonetz, at an elevation of some seven hundred feet above the level of the adjacent country, being the greatest altitude in the government of Olonetz.

"This ridge constitutes the watershed of streams flowing on the one side to the Baltic, and on the other to the White Sea. On the former are narrow lakes, which, with the rivers connecting them or issuing from them, flow into the Onega, while on the latter is the Matkozero, whose waters flowing northward follow the course indicated.

"On the banks of the Matkozero they fell wood for the saw-mills at Povonetz, transporting it by carts across the Mosselgie ridge, the woodmen going further and further into the interior of the forest, in consequence of the exhaustion of the woods near to the saw-mill.

"Having crossed the ridge, I found myself in a country manifesting all the characteristics of a northern land. I got into a boat again, and went by the river some ten versts to the village Telekin situated on a river or lake of the same name—I say river or lake because it is difficult sometimes to designate precisely what is seen by the one name or the other, or to tell at what point it ceases to be one or the other, and to take the different character where it should be called a narrow lake and where it should be designated a broad river.

"The general character of the waters in these regions is the following:—Picture to yourself a comparatively small lake, having a flow barely noticeable in some one direction. In the direction of this flow the water becomes perceptibly narrower, and the shores get higher, and the water takes the form of a river, distinguishable from the lake above by being narrower and having a greater current, or it becomes a strong rapid, by which the waters flow into a large expanded lake, which serves as a reservoir for the waters of the surrounding neighbourhood. Such are the general characteristics of all the small expanses of waters in this region.

"All the rivers and rivulets here have a great many rapids throughout their course. For example, the river Vuigozero, which in a course of 100 versts, 66 miles, from its leaving the lake of that name, to its flow into the White Sea has seventeen rapids. The fall of the river through these rapids is 272 feet. In consequence of these rapids all navigation of the river is out of the question. Only timber is floated down these rivers and their confluents in spring—and this notwithstanding the stones with which the beds are filled, and other obstacles. From the Vuigozero Lake I went 40 versts, 27 miles, by the river Telekin to its embouchure.

"The Vuigozero or Vuigor Lake is one the largest lakes in the district of Povonetz. It is 60 versts or 40 miles long, and 30 versts or 20 miles broad. It is throughout its whole extent studded with islands, which according to the idea prevailing in the locality are equal in number to the days of the year. Some of these have an area of 50 square versts. Many are covered with woods, but uninhabited and unsurveyed, so that their contents are unknown ; many of them find no place on the map, and their area is considered as lake, though some of them have good available soil, or are covered with valuable forests of pine.

"On the shore of the Vuigozero is the Vuigozero Podost, the most southern station in the government of Olonetz for village administration, and this uninviting spot must be for a time my place of residence."

Mr. Judrae states what his duties were, and communicates some valuable information relative to forest operations there and in similar localities, all of which may be afterwards given in detail. Here it is this journey and the aspect and condition of the country as seen by him alone which engages our attention. Of this he thus resumes details :—

"My duties on the forest estate of Vuig being finished, on the 9th of September I left this place to go further to the north and the north-east, to that part of the Povonetz district inhabited by the Corrells or Karrells ; I had to go by boat on the Vuigozero. We had a favourable wind, and the well-filled sails carried the small boat along with great rapidity.

"I might have proceeded directly to my destination, but I could not deny myself the pleasure, being there, of visiting the village of Voitzi, situated at the northern extremity of Vuigozero, 100 versts distant from the White Sea, and now in the government of Archangel. This village is well known as the site of a gold mine which is now a thing of the past. Gold was discovered there in 1735 by a peasant, Tarass Antonoff. Mr. Poushkaroff, in describing the government of Archangel, says that at Voitzi, quartz on being crushed and washed yields 7½ zolotnicks of gold for every 150 pounds.*

"The working of the mine at Voitzi was discontinued in 1783. In 1827 gold was discovered on the banks of the river Vuig, and in the course of the present century investigations have been made several

* 96 zolotnicks = 1 lb. Russia, 40 lbs. Russ. or 36 lbs. avoirdupois = 1 pood.—J. C. B.

times by private parties, but they have not proved successful in un-earthing any stores of the precious metal, and at the present time there remain only here and there pits and buildings in which the workmen of a former day were lodged. The traditions of the district give in a thousand different forms pictures of the prosperity enjoyed by the peasants in those times. I had to pay somewhat dearly for the gratification of my curiosity to see the old mine. I had to go on foot 35 versts, 22 miles, to the nearest Karrell village, situated on the edge of the forest estate of Padan; and my walk was the more unpleasant because the road, or, speaking more correctly, the path, according to local phraseology, founded on the topographical con-dition of the ground, lay *across* the *earth,* and did not go *with* the *earth.* From the northern part of the district of Povonetz on to the shores of the White Sea the ground lies in parallel rows of ridges or linear hillocks, with hollows consisting sometimes of peat bog lying between. The ridges are narrow and long, as are likewise the bogs by which they are separated. They run north and south, conse-quently for the traveller in either of these directions the path lies along the summit of the ridge, and according to local phrase he goes *with the earth,* and it is more easy to do so; but if he travels east or west he must walk across the bogs and ridges ; and as the crests are about a verst apart, this has to be done in every verst of his journey.

"The Padan forest estate lies to the west of Vuigozero, and covers an area of 570,000 *desatins.** All that has been said of the forests of Vuig might be reaffirmed of these, the only difference being that these show a decided preponderance of pines over the number of firs, especially in the northern parts. In general, the nearer we approach to the sea the more rarely do we meet with fir, and at last this tree disappears entirely. With regard to the quality of the pine I can state as the result of my personal observation, that within certain limits the quality of the wood does not depend on the latitude, but is in direct relation to the quality of the soil. In the Vuig forests they are met with first in the middle and northern parts of the Padan forest estate.

"With the development of forest operations in this district the Padan forest will acquire much importance in consequence of the number of navigable streams existing in the White Sea basin. In the southern part of this forest estate there is the Segozera, second in size only to the Onega ; further to the north is the Lake Ondazero, through

* A *desatin* = 40 × 60 or 2,400 Russian square fathoms of 7 English feet. An English acre is 0·37041 of a *desatin,* which makes a *desatin* = 2·69972 English acres. A *desatin* = 4·2789 Prussian morgens. A verst is equal to two-thirds of an English mile.—J. C. B.

which flows the river Onda, one of the tributaries or confluents of the Vuig, constituting the boundary between the governments of Olonetz and Archangel.

"The population of this district is exclusively Correll or Karrell, and the whole country bears the name of Karelia. If a line be drawn on the map from the fall of the Vuig into the White Sea to the town of Povonetz, the whole country to the west of this line lying between it and Finland is Karelia; the Svir may be considered the southern boundary of this, and 65° N. Lat. its northern limits; while the lands bordering on this—the banks of the rivers (the Svir and the Vuig) and of the lakes (Ladoga, Onega, Vuigozero) and the shores of the White Sea—are peopled by Novogorod tribes. Karelia, forming a separate and definite district, has both physical and economic conditions of its own, and peopled by a distinct people who have retained their own language and customs, its invaluable forests have remained for most people a *terra incognita*. These differences, combined with the un-communicativeness of the people, and their rough manners and poverty, which have become proverbial, have impeded the acquisition of information about the country.

"Want compels the Karrells to leave their villages almost every year, and multitudes of them go wandering about towns and villages, begging food in the name of Christ. The condition of the Karrells in the vicinity of Povonetz and of Kem is somewhat better; then through contact with the more energetic Russians they have almost lost their national characteristics.

"On the 1st of October I crossed the frontier of the government of Olonetz, and in three days I was on the shores of the White Sea."

J. AND W. RIDER, PRINTERS, LONDON.

Crown 4to., cloth gilt, 25s., post free.

A TREATISE

ON THE CONSTRUCTION AND OPERATION OF

WOOD-WORKING MACHINES, including a History of the Origin and Progress and Manufacture of Wood-working Machinery. By J. RICHARDS, Mechanical Engineer.

This work contains twenty-five folding plates, and nearly one hundred full-page Illustrations of English, French, and American Wood-working Machines in modern use, selected from the designs of prominent Engineers. The Engravings are finely executed, consisting mainly of true elevations.

"With the exception of a very few imperfect articles in encyclopædias, two or three papers read before scientific societies, and the description of special machines contained in patent specifications, and in the pages of technical journals, such as our own, the subject has been left untouched; and, until the appearance of the book now before us, there was, so far as we are aware, none that could be referred to for information as to how wood machines should be constructed and managed."— *Engineering.*

Cloth gilt, price 3s. 6d.; post free, 3s. 9d.

WOOD CONVERSION by MACHINERY. —By JOHN RICHARDS, M.E., Author of a "Treatise on the Construction of Wood-working Machines," "The Operator's Handbook of Wood-working Machinery for Practical Workmen," "Workshop Manipulation," &c., &c.

ABRIDGED CONTENTS.

Economic Effect of Machine Tools for Wood Conversion—The Inventions of Sir Samuel Bentham—Veneer Cutting—The Rotary Wedge—Saw Guides—Lumber Guides—Curvilinear Sawing—Wave Moulding—Saw Grinding—Wood Bending—Rotary Tools for Wood Cutting—Wood Boring, Patent Monopoly in Wood Conversion—Factory Buildings for Wood Manufacturers—Foundations for Reciprocating Machines—Foundations for Rotary Machines—The Effect produced by Foundations and Frames—The arrangement of Woodworking Factories—The arrangement of Saw Mills—Sawing Frames and their Supports—Different Systems of Manufacture—The Separation of Sawing and Finishing Processes—Engineering Skill in Constructing Wood Factories—Saws and Sawing—Abrading Saws—Saws for Wood—The Form of Teeth—The Operation of Circular Saws—Reciprocating Saws for Deal Splitting—Band Saws for Deal Splitting—Shaping Wood—Longitudinal and Transverse Planing—Planing Carriages—Presenting Wood by Rotation, &c., &c.

Crown 8vo., cloth, 6s., post free.

A TREATISE ON THE ARRANGEMENT, CARE, AND OPERATION OF WOOD-WORKING FACTORIES AND MACHINERY; forming a complete

OPERATOR'S HANDBOOK OF WOOD-WORKING MACHINERY for Practical Workmen. By J. RICHARDS, M.E.

Price One Shilling.

AFTER THE TURTLE. — Thirty-one Years' MINISTERIAL POLICY as set forth at LORD MAYOR'S DAY BANQUETS from 1848 to 1878, collected by R. SEYD, F.S.S. London : J. & W. RIDER, 14, Bartholomew Close, E.C. Manchester: J. BEYWOOD, 141, Deansgate ; and of all Booksellers and Railway Bookstalls.

Price 6d., post free.

NEW & IMPROVED POCKET TABLES, showing equivalent prices of all SIZES DEALS and BATTENS at per foot run and cube, to PETERSBURG STANDARD.—By W. BRYAN.

NINTH EDITION, just published, Price 4d., post free.

THE TIMBER MERCHANT'S POCKET COMPANION. By CHARLES GANE.

Bound in stiff cloth, for the pocket, price 1s. 6d., post free.

NEW FLOORING TABLE, giving the values of Foreign prepared Flooring Boards by the customary square from 3d. to 10s., compared with the Nominal Petersburg Standard, by which they are usually imported, &c.

Price 2s., post free, 2s. 1d., Second Edition, carefully Revised and Enlarged.

THE TIMBER MERCHANT AND BUILDER'S VADE MECUM. By GEORGE BOUSFIELD, 26, Leonard Street, Hull. Containing nearly 3,000 Calculations, and 400 Marks and Qualities of Deals, besides a copious Wages Table, and other miscellaneous information for the use of Timber Importers, Merchants, Contractors, Saw-mill Proprietors, Builders, Joiners, Cabinet-makers, Wheelwrights, Tallymen, and Wood Dealers generally.

OPINIONS OF THE PRESS.

"We cordially recommend to all the Trade this most useful compilation as the best of the kind we have seen, and no one meddling in Timber should be without it."— *Timber Trades Journal.*

"We have pleasure in commending this work as a cheap and simple series of Tables, supplying a want previously felt by all in the Trade."—*Building News.*

"This work quite exhausts the subject it deals with, and will save considerable labour to all who have to do with wood, is well got up, and being portable in form and remarkably inexpensive it should have a good sale."—*Builder's Weekly Reporter.*

"A very useful set of tables done up in a conveniently small form, and will be found of great service to Timber Merchants, &c."—*Furniture Gazette.*

A HIGH-CLASS WEEKLY JOURNAL, PRICE SIXPENCE.

THE LAND AGENTS' RECORD and REAL PROPERTY GUIDE. The following are the principal features :—A complete list of all forthcoming Property Sales, compiled from exclusive information, and containing particulars of every important property in the market throughout the country.—Brief details of Important Sales with prices realized.—Important Compensation Awards, and decisions affecting Landed Interests, with Verbatim Reports where necessary.—Notes on disputed legal points of interest to the profession, edited by Thomas Bateman Napier, Esq., solicitor, Clifford's Inn Prizeman, Scott Scholar, and Conveyancing Gold Medallist.—Short, pithy Editorial Notes on professional topics.—Special Articles of high literary merit, descriptive of properties in the market of unique importance.—The proceedings at the fortnightly meetings of the members of the Institution of Surveyors will be regularly noticed.—Plans, drawings, and elevations inserted as advertisements. Terms of subscriptions, in advance—£1 8s. per annum.

Published on the 1st of every month

THE

JOURNAL

OF

FORESTRY

AND ESTATES MANAGEMENT.

Devoted to the interests of Arboriculture in its Scientific, Practical, Economic, and Ornamental Aspects, and the General Management of Estates.

SINGLE COPY, ONE SHILLING. By Post, 1s. 1½d.
ANNUAL SUBSCRIPTION (post free), TWELVE SHILLINGS.
Specially Reduced Terms to Assistant Foresters, &c.

The special objects of this Journal are to encourage the Study of Arboriculture in its many aspects; to stimulate increased interest amongst landed proprietors in planting trees; to be the Forester's newspaper, giving regularly all items of interest, such as new appointments, promotions, plantings, recent sales of wood, and prices realized in various districts, &c., &c., and generally to supply the long-felt want of a medium of inter-communication between

Landed Proprietors, Foresters, Factors, Stewards, Estate Agents, Nurserymen, Wood Dealers,

and all the various classes interested in Arboriculture and the practical management of Large Estates throughout the United Kingdom.

The List of Contributors includes some of the best known scientific and practical writers of the day, both on Forestry and Estates Management; while by means of "THE EDITOR's BOX" highly interesting and instructive discussions on practical topics are always being carried on; and in the department of "NOTES AND QUERIES" much useful information is diffused by the answering of inquiries.

LONDON:

J. AND W. RIDER, 14, BARTHOLOMEW CLOSE, E.C.

EDINBURGH: DOUGLAS AND FOULIS, 9, SOUTH CASTLE STREET.
DUBLIN: W. H. SMITH AND SON, MIDDLE ABBEY STREET.
NEW YORK: D. VAN NOSTRAND, 23, MURRAY STREET.